INDIAN MATHEMATICS
An Introduction

INDIAN MATHEMATICS
An Introduction

ASHOK JHUNJHUNWALA
Indian Institute of Technology
Madras

PUBLISHING FOR ONE WORLD
NEW AGE INTERNATIONAL (P) LIMITED, PUBLISHERS
New Delhi • Bangalore • Chennai • Cochin • Guwahati • Hyderabad
Jalandhar • Kolkata • Lucknow • Mumbai • Ranchi
Visit us at **www.newagepublishers.com**

Published by New Age International (P) Ltd., Publishers
First Edition : 1993
Reprint: 2009

Branches:

- 36, Malikarjuna Temple Street, Opp. ICWA, Basavanagudi, **Bangalo** © (080) 26677815
- 26, Damodaran Street, T. Nagar, **Chennai**. © (044) 24353401
- CE-39/1016, Carrier Station Road, Near Paulson Park Hotel, Ernakulam South,**Cochin** (0484) 4051303
- Hemsen Complex, Mohd. Shah Road, Paltan Bazar, Near Starline Hotel, **Guwahati**. © (03 2543669
- No. 105, 1st Floor, Madhiray Kaveri Tower, 3-2-19, Azam Jahi Road, Nimboliadda, **Hyderab** © (040) 24652456
- RDB Chambers (Formerly Lotus Cinema) 106A, Ist Floor, S.N. Banerjee Road, **Kolkata** (033) 22275247
- 16-A, Jopling Road, **Lucknow**. © (0522) 2209578
- 142C, Victor House, Ground Floor, N.M. Joshi Marg, Lower Parel, **Mumb** © (022) 24927869
- 22, Golden House, Daryaganj, **New Delhi**. © (011) 23262370, 23262368

ISBN(10) : 81-224-0573-8
ISBN (13): 978-81-224-0573-6

Rs. 100.00

C-09-09-4050

Printed in India at Ramprintograph, Delhi.

PUBLISHING FOR ONE WORLD
NEW AGE INTERNATIONAL (P) LIMITED, PUBLISHERS
4835/24, Ansari Road, Daryaganj, New Delhi-110002
Visit us at **www.newagepublishers.com**

Preface

India has a rich tradition in mathematics. It is however believed that all that was known in India in the past is encompassed today by modern mathematics. I was educated with this understanding. It was therefore a big surprise when in the last few years, I encountered techniques from the Indian mathematical tradition that are amazingly simple and powerful. Besides, several of these techniques were being used by our artisans and businessmen till very recently. I, on the other hand was unaware of most of these methods inspite of my education being heavily based on mathematics. I started talking about them, and in the last six to eight months gave a number of formal and informal talks on this subject. The enthusiastic response that I received, not only from students and teachers, but also men and women from different walks of life, motivated me to put together some of these techniques in this book. Most of what is contained here can be, and perhaps should be, taught in schools from class IV to IX. However the book is written with a larger audience in mind because I feel that more people need to know these techniques and I believe they will enjoy them.

It perhaps needs to be emphasised that the material covered here represents but a small part of Indian mathematics. Even when we do talk in admiring tones about our mathematical heritage, very rarely are the techniques and procedures examined carefully and elaborated clearly for a modern readership. I have made an attempt to rectify this lacuna in the present work. I would also like to point that these techniques would probably have interesting implications from the point of view of their machine-implementation on modern computers. While some people are carrying out this enquiry, the present work does not deal with this aspect.

It is probably worthwhile to dwell a little on the circumstances that eventually led to my writing this book. Foremost amongst these is my close association with the PPST Foundation which has been attempting to re-examine our S & T heritage from the point of view of our contemporary knowledge and concerns. This effort, along with those of many others, has contributed to creating an awareness that our S & T heritage in diverse domains may not merely be a matter of a proud past, but may have significant contributions to make to our present-day endeavours and pursuits. It is the sharing of this conviction that made me take a serious look at traditional mathematical techniques.

It also helped to be a part of the nationwide network involved in the DST-funded project *Foundations and Methodology of Theoretical Sciences in Indian Tradition*. What ultimately gave the necessary motivation and impetus to produce this book at this time, is the holding of the *Congress of Traditional Sciences and Technologies of India* at Bombay. It is hoped that this book, along with similar books and monographs on other themes pertaining to our traditional practices and knowledge, being brought out in the context of the Congress, would create enough interest in our S & T community to initiate systematic and exhaustive examination of all aspects of our heritage in this domain.

A large number of colleagues, students and friends have helped me in different ways in bringing this work to the present shape. I gratefully acknowledge this debt with joy. I thank Dr. Vidyadhar Kudalkar for helping me with typesetting this manuscript. I warmly acknowledge Ms. Renuka, Ms. Anandhi and Ms. Jeyanthi for their assistance in preparing the manuscript and for their patience and commitment.

Ashok Jhunjhunwala

Contents

Contents

Chapter 1

Introduction

About 10 years ago, I had called a carpenter to my home to make a bookshelf. I wanted a special book-shelf with partitions of various sizes. I explained my requirements to him. He took measurements of the walls on which the book-shelf was to be fitted. He noted the size for each of the shelves and partitions. On completion of this, I requested him to compute the amount of wood required and how much the book-shelf would cost. He started the calculations. Simultaneously, I started the same calculations for cross-verification. He completed his calculations in about four minutes. It took me well over ten minutes to finish the calculations and find out that he was right. I was totally taken aback! I always thought I could carry out calculations faster than others. And this carpenter had probably never ever gone to a primary school.

I spent the next half hour prodding him to tell me how he carried out his calculations. What emerged was a simple amazing method. I will illustrate this with an example :

Example : Compute the area of a wood piece of 5 ft 1 in length and 3 ft 5 in breadth.

The method I used for carrying out the calculation was the normal one.

5 ft 1 in = 61 in
3 ft 5 in = 41 in

61×41 = 2501 sqin

2501 sqin is equal to

```
144 )  2 5 0 1 (17
       1 4 4
       -------
       1 0 6 1
       1 0 0 8
       -------
           5 3
```

The answer is 17 sqft 53 sqin.

The method used by the carpenter was much simpler. It involves the following,

(i) Multiply the inches-part of the length and breadth respectively to obtain square inches and write it under the inches column as follows : $1 \times 5 = 5$

5 ft	1 in
3 ft	5 in
	5 sqin

(ii) Multiply the feet-part of the length and breadth respectively to obtain the square feet and write it under the feet column : $3 \times 5 = 15$

5 ft	1 in
3 ft	5 in
15 sqft	5 sqin

(iii) Cross-multiply the feet-part of the length with the inch-part of the breadth, and the feet-part of the breadth with the inch-part of the length. Add the two products and write it in between: $5 \times 5 + 3 \times 1 = 28$

5ft		1 in
3ft		5 in
15 sqft	28	5 sqin

(iv) The result so obtained 28, is divided by 12. The quotient, 2, is added to the square-feet-part of the result obtained earlier. The remainder, 4, is multiplied by 12 and added to the square-inches-part obtained earlier.

5 ft		1 in
3 ft		5 in
15 sqft	28	5 sqin
2 ⟵	$\frac{28}{12}$ ⟶	4 × 12
quotient		remainder
17 sqft		53 sqin

The result obtained is what we desire.

The procedure is very simple and with a little practice can be carried out mentally. Let me illustrate with another example:

Example : Multiply 7 ft 8 in with 3 ft 9 in.

7 ft		8 in
3 ft		9 in
21	87	72
7		3 × 12
28 sqft		108 sqin

Try computing it with the normal method. We would of course get the same result but with much greater effort.

I would like to point out here that the result of the cross–product in the carpenter's method has units of feet × inch. When we divide this by 12 inches per foot, the inches cancel out and we get the square-feet-part. The remainder, 3 feet × inches, we multiply by 12 inches per foot to obtain the square-inch portion. Thus 87 ft × in becomes 7 sqft 3 ft × in or 7 sqft 36 sqin in the second example.

Occasionally we may have to do a little more work. Let me illustrate this with a third example :

Example : Multiply 4 ft 9 in with 3 ft 8 in.

4 ft		9 in
3 ft		8 in
12	59	72
4		11 × 12
16 sqft		204 sqin

This answer of 16 sqft 204 sqin should be written as (16+1) sq ft and (204 – 144) sqin or 17 sqft 60 sqin. That is, whenever the square-inch-part is greater than 144 sq in, the square-feet-part should be incremented by one and the square-inches-part reduced by 144. The calculation, however, could have been carried out a little differently as shown below :

4 ft		9 in
3 ft		8 in
12	59	72
5		−1 × 12
17 sqft		60 sqin

Here, noticing that the remainder, when we divide 59 by 12 is large and is likely to lead to a square–inch–part greater than 144 sq in, one could instead write 59 ft × in as 5 sqft with a negative remainder of 1 ft × in. To get the final square-inch-part of the result, we have to add (−1) × 12 to 72.

The procedure is not only simple, but only requires the ability to carry out multiplications, divisions and additions of small numbers as compared to the normal method. I can understand why the carpenter could carry out the calculations so effortlessly. It is quite possible that the carpenter would have found the normal method used by most others beyond his capacity. Yet, armed with the knowledge of this simple technique which he learnt from his father (and knowing the multiplication tables well, which he had memorised when he was five years old), he could effectively carry out his trade. What I cannot understand is why I never learned such a simple technique known to our artisans at no time in my school or college education, or even during my teaching career which is heavily based on mathematics.

Before I proceed, let me state that this technique is not specific to feet and inches. It can as easily be applied to other units as well. Let us show this with an example using cm and mm as units.

Example: Multiply 4 cm 9 mm and 3 cm 8 mm.

4 cm		9 mm
3 cm		8 mm
12	59	72
6		−1×10
18 sqcm		62 sqm

Note: Since 10 mm is 1 cm, the division (and multiplication) of the cross–product, 59 cm × mm is carried out by 10 now.

II

I am reminded of another computational technique which I was taught when I was six years old. After school or on holidays, I often used to go and sit in our family-business-office which happened to be in the same premises as our house. I used to sit with my uncles as they maintained the accounts and ledgers. Occasionally, I would be given some additions to perform, and I used to enjoy it. During these times, I also used to be taught some calculational techniques, many of which I do not remember today. But one of the techniques taught by my uncle has stood by me throughout my life. Let me describe it now to you.

Example : Multiply 317 by 437 and check the result.

		Navasesh calculation		
	317	3+1+7 = 11 => 1+1=2		
×	437	4+3+7 = 14	1+4=5	
	2219		10	1+0=1
	951			
	1268			
	138529	1+3+8+5+2+9=28,	2+8=10	1+0=1

The multiplication is carried out by the well-known long multiplication method. What I was taught was a simple method called *Navasesh* (meaning Modulo-Nine) to check if my calculations were correct. The technique involves the following,

(i) Take the first number and add the individual digits of the number to obtain their sum. If the result is more than one digit long, add the individual digits of the result also. Continue in this manner till you finally get a single digit. This digit is the Navasesh or Modulo-Nine of the original number. The Navasesh of 317, designated hereafter as **N**(317) is **N**(3+1+7) or **N**(11) which is equal to 2.

(ii) Similarly take the Navasesh of the second number, 437. The **N**(437) is **N**(4+3+7) or 5.

(iii) Multiply the two Navasesh obtained above. Once again add the individual digits of the result to obtain the final Navasesh as 1.

(iv) Now take the Navasesh of the product of 317 and 437. The Navasesh of 138529 is (1+3+8+5+2+9) which is **N**(28) or 1. Note that the Navasesh of this is same as the Navasesh obtained in (iii).

If the calculation is correct, this match would always take place. In other words,

$$\mathbf{N}[\mathbf{N}(a) \times \mathbf{N}(b)] = \mathbf{N}(a \times b)$$

or in our case,

$$\text{LHS} = \mathbf{N}[\mathbf{N}(317) \times \mathbf{N}(437)] = \mathbf{N}[2 \times 5] = 1 \text{ and}$$

$$\text{RHS} = \mathbf{N}[317 \times 437] = \mathbf{N}[138529] = 1$$

This is a simple check and can be performed to check the correctness of *any multiplication*. A lack of this match implies that an error has been made in calculations.

Let us take another example to illustrate the same principle,

		Navasesh	
	3526	(7)	
×	2425	(4)	**N**(7× 4)=**N**(28) =(1)
	8550550		

As the two Navasesh match, one may assume that the multiplication is correct.

The Navasesh, as its meaning indicates, is the remainder when a number is divided by 9. It is easier, however, to compute this remainder as shown, rather than by divison. It is approriate here to mention a few points about the calculation of Navasesh. Note that 9+1 is 10 whose Navasesh is 1. Similarly Navasesh of 9+2 is 2, Navasesh of 9+3 is 3, Navasesh of 9+4 is 4 and so on. Therefore, the number 9 in the Navasesh calculation is equivalent to 0. It can be dropped at any stage. Also, in Navasesh calculation, an intermediate two digit result, can be reduced to a single digit by adding the two numbers. Thus the Navasesh of 8550550 could be calculated as

8+5=13 which is same as 4
4+5+0+5=14 which is same as 5
5+5+0=10 which is same as 1.

The Navasesh of 8550550 is therefore 1.

A few more examples illustrate the use of Navasesh. (The Navasesh in subsequent examples are written to the right of the number in parenthesis. When two numbers are to be multiplied, their Navasesh are also multiplied and reduced to a single digit and written in parenthesis on the right after a vertical bar.)

```
   1435      (4)                 523     (1)
   3421      (1) | (4)          6245     (8) | (8)
 -------                      -------
 4909135     (4)             3266135     (8)

   536      (5)
    72      (9) | (9)
 -------
 38592      (9)
```

This procedure of using Navasesh to check calculations is not limited to multiplication. Let me illustrate a similar checking procedure for addition :

	527		(5)	
	346		(4)	
+	259	+	(7) | (16)	=> (7)
	1132		(7)	

Note that the Navasesh of individual numbers are added together and its Navasesh is the same as the Navasesh of the sum. The same can be expressed as

$$\mathbf{N}[\mathbf{N}(a) + \mathbf{N}(b) + \mathbf{N}(c)] = \mathbf{N}(a+b+c).$$

This technique also works for subtraction. The only point that need to be remembered is that a negative Navasesh can be converted to a positive Navasesh by adding 9. Thus Navasesh (–5) = 4 and Navasesh (–7) = 2.

Let us illustrate subtraction with an example :

4952	(2)
3527	(8) | (–6) => (3)
1425	(3)

Since the technique for checking is valid for addition, subtraction and multiplication, it must work for any algebraic expression as illustrated below :

$$435 \times 16 + 315 \times 12 = 10740 \qquad (3)$$

$$\text{Navasesh:} \quad (3) \times (7) + (9) \times (3) = (3)$$

The method is simple to use and a lack of match definitely indicates an error. I should however warn here that a correct *match does not guarantee that the result is correct.* A little thought will indicate that when two digits of the correct result are interchanged to give an incorrect result, the Navasesh will be unchanged and will lead to a match. However, when we carry out the calculations and perform the check and find a match, an error will remain undetected only very occasionally.

I learnt this technique when I was six years old and have always used it since then. The technique has always enabled me to check my calculations. It seemed to help me stay ahead in my classes and was also responsible in part for my interest in mathematics. What surprises me is that only a few people know this technique. I remember an incident involving some of my colleagues. We were all correcting some examination papers together in a hall. Finally we divided the papers for tabulation work, which essentially involved large additions and some multiplications. We were supposed to tabulate the results, as well as check our calculations. Two of us completed the work much faster than the others. On enquiring, I found that this colleague also knew Navasesh and had used it. He was taught this technique when he was very young.

I recently found out that Navasesh was used in our family account books for cross-checking. A typical profit and loss account was written as follows :

	Purchase	
(9)	12×0.36 =	4.32
(3)	24×0.62 =	14.88
(9)	36×0.44 =	15.84
(9)	18×0.38 =	6.84
(3)		41.88

	Sale	
(9)	4×0.45 =	1.80
(3)	6×0.50 =	3.00
(4)	8×0.50 =	4.00
(6)	4×0.60 =	2.40
(3)	8×0.60 =	4.80
(9)	6×0.60 =	3.60
(3)	12×0.58 =	6.96
(3)	12×0.64 =	7.68
(6)	8×0.75 =	6.00
(1)	8×0.80 =	6.40
(3)	6×0.80 =	4.80
(2)	8×0.70 =	5.60
(7)		57.04
−(3)	Purchase	− 41.88
(4)	Profit	15.16

The account involves purchase and sale of items. For each entry the Navasesh of the transaction was written in parentheses. This is calculated easily by taking the Navasesh of the quantity and that of the price, multiplying them and taking the Navasesh of the product. Thus Navasesh of 12 × 0.36 is 3 × 9 which is 9. The Navasesh of the total purchases and sales are obtained by adding the Navasesh of the individual transactions. Finally, the purchase-Navasesh is subtracted from the sale-Navasesh. The result must match the Navasesh of the profit. This single Navasesh entry at the end of the page enables one to verify the calculations of the whole page at a glance. It may not be out of place to mention here that the principle of Navasesh is very much like that of parity-check used in modern day communications and computations. The distinctive feature of Navasesh is that it is easy to compute.

III

These simple techniques for computing the area of a rectangle or checking arithemetic operations, have been known and used in India for quite some time. What is important, however, is that they were used not only by experts or pandits, but also by the common people to carry out their trade. While it may be difficult to trace their origins, it is possible that they were developed by people carrying out different trades in accordance with their specific needs. I had never given much thought to whether there are other similar mathematical techniques.

It was therefore again a surprise when about 4 years ago, I came across the book *Vedic Mathematics* by Bharathi Krishna

Tirtha Maharaj [Motilal Banarsidass Publishers Pvt Ltd, Varanasi, 1965]. I was amazed to find several other techniques and algorithms which make arithemetic much simpler and more interesting. Most of these techniques and algorithms follow from a deep understanding of the place value system in mathematics. Such an understanding of the place value system has existed in India for well over a thousand five hundred years and is well illustrated in the works of Aryabhata, Brahmagupta, Mahavira, Sripati, Sridhara, Bhaskaracharya and so on.

Several of the techniques and algorithms described in Bharathi Krishna's book *Vedic Mathematics* have been in use in India for a very long time. It is quite likely that these techniques were widely used by different sections of our people till about 80-100 years ago. It is therefore surprising that these techniques were not noted or picked up by our modern mathematicians and collected together in a school-level book .

This book is written to illustrate some of these techniques and algorithms. It is my belief that an understanding of these at a young age will help a child understand and enjoy mathematics much more; this understanding will enhance creativity in young minds besides developing a healthy respect for our own traditional sciences and technologies.

We had a rich tradition of mathematics in India with excellent work being carried out even as late as early part of 19th century in places such as Kerala. We have barely begun to document and understand these works. This book constitutes only a small, though interesting, part of the Indian mathematical tradition.

Exercises

1. Find the area of a rectangle with the following lengths and breadths using the method used by the carpenter.

a) L = 6 ft 7 in , B = 4 ft 9 in

b) L = 8 ft 4 in , B = 5 ft 11 in

c) L = 8 cm 7 mm, B = 4 cm 9 mm

d) L = 9 cm 1 mm, B = 2 cm 7 mm

2) Perform the following additions, subtractions and multiplications and check using Navasesh.

a)
```
   3 7 2 4
   2 1 4 9
   3 2 7 4
+  2 1 5 9
  ________
  ________
```

b)
```
   3 2 4
   2 1 8
   2 9 4
   3 7 6
+  4 5 2
  ______
  ______
```

c)
```
   4 5 9 3 2
-  3 1 7 9 4
  __________
  __________
```

d)
```
   4 2 7
×  3 4 9
  ______
  ______
```

e)
$$
\begin{array}{r}
7384 \\
\times \quad 48 \\
\hline \\
\hline
\end{array}
$$

f)
$$
\begin{array}{r}
48253 \\
\times \quad 82 \\
\hline \\
\hline
\end{array}
$$

3) Compute the following expressions and verify using Navasesh.

a) $314 \times 4 + 24 \times 18$

b) $71 \times 12 - 31 \times 11$

c) $12 \times 14 + 31 \times 9 + 41 \times 11$

d) $11 \times 7 \times 4 - 41 \times 3$

Chapter 2

Nikhilam - Special Multiplication

We begin with a very simple multiplication technique which is referred to as Nikhilam. It is a special technique applicable only when the numbers to be multiplied have some special characteristics. We begin with the case when both the numbers are less than and close to hundred. For example, let us multiply 95 and 98. Note that 95 is 5 less than hundred and 98 is 2 less than hundred. Let us write the two numbers and their deviations from hundred with a vertical line separating the two as follows :

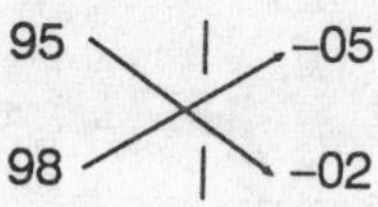

Note that the deviation has two digits. The product of 95 and 98 is obtained by multiplying the numbers to the right of the line (i.e., the deviations 05 and 02), and writing the result at units and tens places of the product. Then, a cross-subtraction is carried out diagonally, either 95 – 02 or 98 – 05 to give 93 which occupies the thousands and hundreds places as shown below :

```
95  |  –05
98  |  –02
----------
93     10
```

You may be surprised to know that 9310 so obtained is the required product. Check it out using Navasesh described in Chapter 1: **N**(95) = 5; **N**(98) = 8 and **N**(5 × 8) = 4. The Navasesh of 9310 is also 4 indicating that our calculation is correct.

Let us illustrate Nikhilam with a few more examples :

```
91  |  –09          89  |  –11
94  |  –06          96  |  –04
----------          ----------
85     54           85     44
```

```
                    Navasesh check
88  |  –12          N(88)   = (7)
91  |  –09          N(91)   = (1)  | (7)
 1
----------
80     08           N(8008) = (7)
```

Note that in the last example the product of the deviations 12 × 9 results as 108. The 08 occupies the tens and units place. The digit one however is carry-over to the hundreds place and has to be added to the cross-subtracted result 88–09 or 79. Verify that the product is correct using Navasesh.

A very similar procedure is applicable when the two numbers are less than and close to 10, or 1000, or 10,000, etc. For numbers close to 10, the numbers on the right of the vertical line are deviations of the number from 10. The procedure for multiplication involves the same two steps, i.e,

(i) multiply the deviations, and

(ii) cross-subtract the diagonal number as shown below:

9	–1
7	–3
6	3

9	–1
9	–1
8	1

8	–2
7	–3
5	6

Similarly for numbers close to 1000, the deviation from 1000 is three digits as follows:

998	–002
980	–020
978	040

997	003
888	–112
885	336

Check the calculation using Navasesh. Note that we write –002 to the right of the vertical line, and not –02 or –2. The number of zeros to be used depends on the base here, 1000 that we are working with. We must also place same number of zeros in the product. For instance, in the multiplication of 998 and 980, 2 and 20 are multiplied to get 40. However, the product has three digits, and therefore 040, to the left of the vertical line.

Similarly for numbers close to 10,000 the multiplication can be carried out as follows :

9997	–0003
9995	–0005
9992	0015

9984	–0016
9950	–0050
9934	0800

It is not strictly required that the numbers always be very close to the base 10 or 100 or 1000 etc. For example, if only one number is very close to the base and the other is not, the calculation is still simple as shown below :

998	–002
598	–402
596	804

This example required us to multiply 402 with 2 and subtract 2 from 598. However, if both numbers are not close to the base, we get situations as follows :

889	–111
942	–058
6	
837	438

84	–16
72	–28
4	
60	48

Note that the first calculation involves multiplication of 111 and 58 which is not simple. It also involves subtraction of 58 from 889 and adding the carried over 6. The second calculation involves multiplication of 16 and 28 and then subtraction of 16 from 72 besides adding the carry. The simplicity of the calculation is lost.

Nikhilam for numbers greater than the base

For numbers close to some power of 10 (i.e., 10 or 100 or 1000, etc), the multiplication can be as simply carried out even if both the numbers are greater than the base. For example, if we wish to multiply 105 with 108, we proceed in a similar manner. To the right of the number we write its difference from 100 with a positive sign (deviation from 100). The numbers on the right of the line are now multiplied as before. The number on the left, however, is obtained by diagonal cross–addition (instead of cross–subtraction) as shown below :

			Navasesh check
105	|	+05	N(6 × 9) = 9
108	|	+08	
113		40	N(11340)=9

Proceeding in a similar manner we can work out the following examples :

12	|	+2		104	|	+04
13	|	+3		114	|	+14
15		6		118		56

108	|	+08		1012	|	+012
124	|	+24		1009	|	+009
1						
133		92		1021		108

Please note that in the third example above when 8 is multiplied by 24, there is a carry over to be taken to the hundreds place.

Nikhilam for numbers above and below the base

Let us now take the numbers to be multiplied to be close to the base (100, 100, 1000 etc), but one number is larger and the other smaller than the base. The Nikhilam procedure for multiplication is again applicable. To the right of the number one must write its deviation from the base with appropriate sign and number of zeros. Once again multiply the two deviations. However, since one of the deviations is positive and the other is negative, the product on the right of the vertical line will be negative. The number is written with a bar on top to indicate that it is a negative number. The left side result is obtained by either cross–addition or cross–subtraction, as shown below :

105 | +05

94 | –06

99 $\overline{30}$ $\quad$ 9900 – 30 = 9870

To obtain the number on the left, we have added 94 and 05. We could have also subtracted 06 from 105 to obtain the same result. The negative part on the right of the line must now be converted to a positive number. $99\overline{30}$ implies the result is *30 short of 9900*, and therefore the result is 9900 – 30 or 9870.

More examples are given below to illustrate the above method :

$$\begin{array}{r|l} 9 & -1 \\ 12 & +2 \\ \hline 11 & \overline{2} = 108 \end{array} \qquad \begin{array}{r|l} 91 & -09 \\ 104 & +04 \\ \hline 95 & \overline{36} = 9464 \end{array}$$

$$\begin{array}{r|l} 1004 & +004 \\ 996 & -004 \\ \hline 100 & \overline{016} = 99984 \end{array}$$

Use of positive and negative number system – Mishrank

The concept *short of 10 by ...* or *short of 100 by ...* or in other words 10's complement or 100's complement, has been used in Indian Mathematics along with the place value system. Indians have often used a combination of positive and negative numbers at units, tens, hundreds and thousands places to make calculations easier. In the examples of Nikhilam multiplication above, we encountered these mixed numbers, also referred to as *Mishrank*. The number $99\overline{30}$ is the Mishrank place–value representation and is equivalent to

$$99\overline{30} = 9 \times 10^3 + 9 \times 10^2 - 3 \times 10^1 - 0 \times 10^0$$

Thus using Mishrank a number like 28 can be written as $3\overline{2}$ or a number 473 can be written as $5\overline{3}3$ or $48\overline{7}$ or $5\overline{27}$. The ability to work with Mishrank helps one to carry out complex calculations easily, as will be seen in subsequent chapters. Mishrank is known to modern mathematicians as the Redundant Number System.

Nikhilam with sub-bases

We have seen that Nikhilam makes multiplication of numbers close to a base very simple. Can we use it for numbers that are not close to 10 or 100 or 1000, but numbers close to say 50 or 250? Noting that 50 is 100 divided by 2 and 250 is 1000 divided by 4, we can use Nikhilam with such sub-bases. If the numbers to be multiplied are close to these sub-bases, the product can be obtained using Nikhilam, as illustrated below.

Let us multiply 43 by 45. Note that both the numbers are close to and less than 50, a sub-base of 100. We find the deviations of these numbers from 50 and write it to the right of a vertical line as before. Since 50 is being considered a sub-base of 100, the deviations will have two digits as shown below :

```
43 | –07
45 | –05
--------
     35
```

The deviation 07 and 05 are then multiplied to obtain the tens and unit place of the result. The diagonal cross-subtraction is also carried out to obtain 43 – 05 or 38. However, this is not straightaway the final result at the hundreds place. Since we are carrying out the calculations using, a sub-base 50, the number 38 obtained above is to be divided by 2 to obtain the correct result for the hundreds place. Thus the product is obtained as follows :

43 | –07

45 | –05

38/2 35 = 1935

Check the calculations using Navasesh. **N**(43) is 7 and **N**(45) is 9. The Navasesh of the product of 7 and 9 is 9. The Navasesh of 1935 is also 9, verifying that the calculation is correct.

Let us illustrate the Nikhilam multiplication using a sub-base with another example :

Base: 50=100/2

57 | +07

42 | – 08

49/2 $\overline{56}$ = 2450 – 56 = 2394

Check the calculations using Navasesh. Note that this multiplication is considerably more difficult. First, since one number is greater than 50 and the other less than 50, the product of the deviations is –56. Further, the diagonal cross-subtraction (or cross-addition) results in 49 and this is to be divided by 2 to give the result for the hundreds place. This yields 24.5 and the final result is obtained by 24.5 × 100 – 56 or 2394.

If the sub-base used is 250, the result of the cross-subtraction/addition is to be divided by 4 to obtain the thousands place of the result; thus,

		Navasesh
243	–007	
247	–003	N(9 × 4) = 9
240/4	021 = 60021	N(60021) = 9

The calculation is verified using Navasesh as shown above.

Nikhilam with multiples of a decimal base

It is somtimes useful to carry out Nikhilam multiplication using bases which are multiples of 10 or 100 or 1000 etc. Let us illustrate with an example using a base of 30 which is 3 times 10.

		Navasesh
26	–4	(8)
28	–2	(1) | (8)
3 × 24	8 = 728	(8)

Since the base 30 is three times the decimal base 10, the deviations are single digit numbers –4 and –2. Their product is 8. The diagonal cross-subtraction yields 24. This needs to be multiplied by 3 to give result for the tens and hundreds places. The Navasesh verification is also shown in this example.

Let us now take another, slightly more complex, example using the base of 30 or (3 × 10).

34	+ 4
38	+ 8
3	
3 × 42	2 = 1292

The product of the deviations, 4 and 8, is 32. Since the decimal base is 10, the digit 3 is the carried over. The diagonal cross-addition yields 42 which is to be multiplied by 3. To this the carry 3 needs to be added to get the tens, hundreds and thousands place of the result. A few more examples, with different multiple bases, illustrate this technique.

Base : 60 = 10 × 6

```
62  |  +2
58  |  -2
-----------------------------
6 × 60    4̄  = 3600 - 4 = 3596
```

Base : 200 = 100 × 2

```
194  |  -06
196  |  -04
-----------------------------
2 × 190    24  = 38024
```

Base : 500 = 100 × 5

```
491  |  -09
493  |  -07
-----------------------------
5 × 484    63  = 242063
```

The last problem above can also be solved taking 500 as a sub-base of 1000 as follows :

491 | −009

493 | −007

484/2 063 = 242063

Note that the deviations in the first approach are only 2 digits while in the latter, they are 3 digits.

As mentioned earlier, Nikhilam multiplication becomes complex as deviations from the chosen base increase. Let us illustrate this with the following example :

Base : 600 = 100 × 6

608 | +08

514 | − 86

−6

6 ×522 $\overline{88}$ = 313200 − 688 = 312512

Check the calculation using Navasesh. **N**(608) is 5 and **N**(514) is 1. Their product is 5. The Navasesh of 312 and 512 is also 5.

The above problem can be carried out using a different base, as follows :

Base : 500 = 1000/2

608 | +108

514 | +014

1

622/2 512 = 312512

Both the calculations are relatively complex, and yet much simpler when compared with the usual long multiplication of 608 with 514.

Nikhilam with three-number multiplication

The multiplication technique using *Nikhilam* described above can be extended to multiplication of three numbers as long as all three numbers are close to the same base. Let us illustrate this with a simple example where all the three numbers to be multiplied, say a, b and c, are slightly larger than 1000. The deviation from the base is written as illustrated earlier. The lowest three digits are obtained by multiplying the three deviations. The next three digits are obtained by multiplying two deviations at a time and adding the results, i.e., d(a) × d(b) + d(b) × d(c) + d(c) × d(a), where d is the deviation. The highest digits are obtained by simply adding/subtracting any of the three numbers with the deviations of the other two as follows :

			Navasesh
1002	+002		N[N(1002)×N(1005)× N(1001)]
1005	+005		= N(3×6×2) = (9)
1001	+001		
1008	017	010	N(1008017010) = (9)
(1002 + 5 + 1)	(2×5 + 5×1 + 1×2)	(2×5×1)	

Once again the Navasesh is used for verification. The Navasesh of the three numbers are 3, 6 and 2 and the Navasesh of their product is 9. The Navasesh of the result, 1008017010 is also, 9.

Let us next consider a more complex example, using the base 100.

97		–03
95		–05
96		–04
88	47	$\overline{60}$ = 884640

The deviations –3, –5 and –4 are multiplied to give 60, but with a minus sign (because all three deviations are negative). The middle two digits are obtained as (–3) × (–5) + (–5) × (–4) + (–4) × (–3). The highest two digits are obtained as 97–5–4. The *Mishrank* result $8847\overline{60}$ is then converted to a non-Mishrank number as follows :

$$8847\overline{60} = 884700 - 60 = 884640.$$

When the numbers to be multiplied are above and below a base, the calculation becomes more complex as shown below :

1001		+001
998		– 002
1002		+002
1001	$\overline{004}$	$\overline{004}$ = 1000995996

The lowest three digits are obtained as (+1) × (–2) × (+2) or $\overline{004}$. The next three digits are obtained as (+1) × (–2) + (–2) × (+2) + (2) × (1). The highest three digits are obtained as 1001 – 2 + 2. The *Mishrank* is then converted as follows :

$$1001\overline{004004} = 1001000000 - 004004 = 1000995996$$

Three-number Nikhilam multiplication using sub-base or multiple of decimal base

The above method of multiplying three numbers can be extended to the cases where the base used is either a sub-multiple or multiple of a decimal base. Let us illustrate the procedures with some examples.

Example: Base 50 = 100/2.

		Navasesh verification
51	+01	(6)
53	+03	(8)
52	+02	(7) | (3)
56/4	11/2 06 = 140556	(3)

Example: Base 60 = 10 × 6.

		Navasesh verification
61	+1	(7)
62	+2	(8)
63	+3	(9) | (9)
6×6×66	6×11 6	

= 237600 + 660 + 6 = 238266 (9)

Squaring using Nikhilam

It should be obvious that the multiplication technique using Nikhilam could be used to obtain squares of a number close to any base. Thus for example, the square of 97 can be obtained as follows :

Base 100 :

97	–03		Navasesh
97	–03		$N[N(97) \times N(97)] = 4$
94	09		$N(9409) = 4$

The deviation from the base is simply squared to obtain the lower digits, the addition/subtraction of the deviation from the number to be squared gives the higher digits. Noting this, we can straight away write the squares of the following numbers.

Base 100:	$91^2 = (91–09) \mid 81$	$= 8281$
	$87^2 = (87–13) \mid {}_169$	$= 7569$
	$83^2 = (83–17) \mid {}_289$	$= 6889$
Base 10:	$17^2 = (17+7) \mid {}_49$	$= 289$
	$13^2 = (13+3) \mid 9$	$= 169$
Base 10,000:	$9992^2 = (9992–0008) \mid 0064$	$= 99840064$

Verification of the squares using Navasesh is also simple. For example, **N**(9992) = 2 and that of its square is 4. One can check that the Navasesh of the result 99840064 is also 4.

Special use of Nikhilam

A corollary of Nikhilam is useful to multiply two numbers provided they satisfy the following two conditions,

(i) units digits of the two numbers add to ten

(ii) higher digits are the same, say equal to n.

The product of the two numbers can be obtained by first multiplying the units digits to obtain the lower two digits of the result. The higher digits, n is then multiplied by n+1 to form the higher digits of the result. Let us illustrate this with an example.

Example: Multiply 27 and 23.

```
                     Navasesh
     2 | 7            (9)
  ×  2 | 3            (5) | (9)
  --------
     6 |21            (9)
```

Note that the units digits of 27 and 23 add to 10 and the higher digits are the same. The lower two digits of the result are obtained by multiplying 7 and 3. The higher digit is obtained as 2 × 3 = 6. The verification of the calculation using Navasesh is also shown.

Several other examples are worked out below to illustrate the principle :

```
     7   3          6   2          10   1
  ×  7   7       ×  6   8       ×  10   9
  --------        -------        --------
    56|21          42|16          110|09


     99     8       59     8
  ×  99     2    ×  59     2
  -----------    ---------
   9900|16        3540|16
```

The above technique can also be used to find the square of any number ending with the digit 5. Since both the properties listed above are satisfied in such a case, the result can be obtained in a straight forward manner. The lower two digits will always be 5 × 5 or 25. The higher digits can be obtained by computing n(n+1) where n is the number formed by the higher digits of the number to be squared. This is illustrated as follows :

$$15^2 = 1\times(1+1)|25 = 225$$
$$85^2 = 8\times(8+1)|25 = 7225$$
$$105^2 = 10\times(10+1)|25 = 11025$$
$$185^2 = 18\times(18+1)|25 = 34225$$
$$255^2 = 25\times(25+1)|25 = 65025$$

Check these calculations using Navasesh.

Exercises

1. Perform the following multiplications using Nikhilam and a base of 10 or 100 or 1000 and check your results using Navasesh.

a) 14 × 12	b) 14 × 9	c) 108 × 104
d) 116 × 109	c) 94 × 83	f) 105 × 98
g) 112 × 84	h) 987 × 991	i) 1008 × 1014
j) 894 × 989	k) 1014 × 9 / 8	

2. Perform the following multiplications using appropriate sub-bases or multiples of decimal places and check your results using Navasesh.

a) $\begin{array}{r} 41 \\ \times\ 47 \\ \hline \\ \hline \end{array}$ b) $\begin{array}{r} 57 \\ \times\ 59 \\ \hline \\ \hline \end{array}$ c) $\begin{array}{r} 41 \\ \times\ 53 \\ \hline \\ \hline \end{array}$

d) $\begin{array}{r} 40 \\ \times\ 47 \\ \hline \\ \hline \end{array}$ e) $\begin{array}{r} 237 \\ \times\ 241 \\ \hline \\ \hline \end{array}$ f) $\begin{array}{r} 257 \\ \times\ 259 \\ \hline \\ \hline \end{array}$

g) $\begin{array}{r} 34 \\ \times\ 31 \\ \hline \\ \hline \end{array}$ h) $\begin{array}{r} 242 \\ \times\ 256 \\ \hline \\ \hline \end{array}$ i) $\begin{array}{r} 72 \\ \times\ 74 \\ \hline \\ \hline \end{array}$

j) $\begin{array}{r} 491 \\ \times\ 508 \\ \hline \\ \hline \end{array}$

3. Perform the following three–number mulitplications using Nikhilam and check yor results using Navasesh.

a) 985 × 991 × 994

b) 1008 × 1002 × 1014

c) 984 × 1008 × 996

d) 98 × 102 × 97

e) 54 × 52 × 55

f) 254 × 257 × 252

4. Find the square of the following numbers using Nikhilam and check your results using Navasesh.

a) 94	f) 35
b) 88	g) 115
c) 984	h) 75
d) 109	i) 165
e) 1018	j) 245

5. Perform the following multiplications using Nikhilam and check your results using Navasesh.

a) 37×33
b) 52×58
c) 84×86
d) 113×117
e) 141×149

Chapter 3

Multiplication Using Urdhva Tiryaka

The multiplication techniques described in the previous chapter belong to special cases as they were applicable only when the numbers to be multiplied satisfy certain conditions. What we are going to describe in this chapter is a very general multiplication technique useful in all cases. The technique is based on a principle called *Urdhva Tiryaka* and primarily involves cross-multiplication. This technique was developed in India before the 8th century and is based on a deep understanding of the place-value system of representing numbers.

Let me illustrate the technique with an example where we multiply 534 with 463. We write down the first number as it is, but we reverse the second number (463 is reversed as 364) and write it below the first number with the right-most digit of the first over the left-most digit of the second as shown below :

```
5 3 4              4 × 3 = 12
    3 6 4

   1
-------
    2
```

The first digit of the result is obtained by multiplying the vertically overlapping numbers 4 and 3. The product is 12; the units digit 2 is written as the units digit of the result and 1 is the carried over to the second digit. 364 is then shifted left by a digit. The vertically overlapping numbers are multiplied and the products added. Any carry from the previous operation is added to this to obtain the hundreds digit of the result.

```
5 3 4              3 × 3+4 ×6 = 33
  3 6 4            33 + 1 = 34

3 1
------
  4 2
```

Thus the overlapping digits 3 and 3 as well as 4 and 6 are multiplied and the two products are added to give 33. The addition of carry 1 gives 4 as the second digit of the result, and a carry of 3. The process is repeated by again shifting 364 to the left as shown below :

```
  5 3 4            5 × 3+3 × 6+ 4×4= 49
  3 6 4            49 + 3 = 52

5 3 1
-------
  2 4 2
```

When the overlapping digits are multiplied and added, we get 49, to which we add the carry over 3. The third digit of the product is therefore 2 and 5 is carried over to the fourth digit.

The number 364 is again shifted left and the same operation is repeated as shown below :

```
     5 3 4              5 × 6+3 × 4 =  42
   3 6 4                42 + 5 = 47

 4 5  3 1
 ---------
   7 2 4 2
```

The addition of the products of overlapping digits gives 42 and to this we add the carry 5 to give 7 as the fourth digit of the result, and 4 as the carry. Once again, we shift 364 to the left and repeat the operation.

```
       5 3 4            5 × 4 =20
   3 6 4                20 + 4 = 24

     4 5 3 1
 -----------
 2 4 7 2 4 2
```

As only 5 and 4 are vertically overlapping their product yields 20. The carry 4 is added to give 24. The fifth digit of the result is thus 4, and 2 will be carried over to the sixth digit. Once again shifting 364 to the left by a digit results in no overlap and therefore the sixth digit of the result is only the carry 2. Thus, 534 × 463 = 247242. To check the same using Navasesh we note that **N**(534) = 3 and **N**(463) = 4, and the Navasesh of their product is 3. We also note that **N**(247242) = 3, thus checking the result.

Thus the *Urdhva Tiryaka* multiplication technique involves reversing of one of the numbers and sliding it from right to left below the second number. At each stage, the overlapping digits are multiplied and added to give the succeeding digits of the result. This operation in modern mathematical terminology is called *convolution*. Convolution is widely used in linear system theory by electrical engineers, though it is applied more in the context of continuous functions. Even to modern mathematicians, it is not widely known that convolution can also be used to multiply numbers easily.

Let us take another example to illustrate the technique.

Example : Multiply 523 × 231.

```
523
  132
_____
  3
```

3 × 1 = 3

```
 523
  132
1
______
  13
```

3 × 3 + 2 × 1 = 11

```
  523
  132
11
______
  813
```

5 × 1+2×3+3× 2=17

17 + 1 = 18

```
   523
  132
211
_______
  0813
```

5 × 3 + 3 × 2 = 19

19 + 1 = 20

```
   5 2 3          5 × 2 = 10
 1 3 2            10 + 2 = 12

 2 1 1
----------
1 2 0 8 1 3
```

Check that 523 × 231 =120813 using Navasesh. Urdhva Tiryaka is very simple and can be carried out mentally by just visualising the second number sliding below the first. It only requires some practice. We will work out several examples to illustrate this. Try working out the results mentally.

Example : Multiply 5234 by 2324.

```
        5 2 3 4                 4 × 4 = 16
              ←— 4 2 3 2        4 × 3 + 2 × 4 = 20
                                4×2 + 2×3 + 3×4 = 26
  2 2  4 2 2 1                  4×5+2×2+3×3+2×4 = 41
-----------------               2×5 + 3×2 + 2×3 = 22
1 2 1 6 3 8 1 6                 3 × 5 + 2 × 2 = 19
                                2 × 5 = 10
```

The result is 12163816 whose Navasesh is 1. Note that **N**(5234) = 5 and **N**(2324) = 2 and the Navasesh of their product is also 1. The working at each step is shown on the right hand side. With a little practice, this can be carried out mentally.

Example : Multiply 61262 by 41.

```
      6 1 2 6 2           1 × 2 = 2
             ←— 1 4       1 × 6 + 4 × 2 = 14

    1 1 2 1               1 × 2 + 4 × 6 = 26
---------------           1 × 1 + 4 × 2 = 9
  2 5 1 1 7 4 2           1 × 6 + 4 × 1 = 10
---------------           4 × 6 = 24
```

The result is 2511742. Check the same using Navasesh.

Example : Multiply 52374 by 31231.

```
           5 2 3 7 4
                   ←— 1 3 2 1 3

  1 2 4 3  3 3  1
 ___________________
 1 6 3 5 6 9 2 3 9 4
 ___________________
```

Note that **N**(52374) = 3 and **N**(31231) = 1 and **N**(1635692394) = 3, thus checking the calculation. The calculation has been carried out mentally. It would be very laborious to carry out the multiplication using the long multiplication technique.

The ability to carry out such calculations fast, of course, depends on one's ability to mentally carry out arithmetic operations. It requires that the person is well versed with the multipliation tables (upto 9) and is able to carry out two–digit additions mentally. These skills used to be emphasised in primary schools in India even twenty years ago. Armed with such abilities a 5–digit by 5-digit multiplication can be carried out at amazing speeds.

Use of Mishrank with Urdhva Tiryaka

You may have noticed that most of the multiplication examples that we have considered so far involved mostly small digits (smaller than 5 rather than 7, 8 or 9). The method is also applicable when numbers with larger digits are involved. The only problem is that the product of large digits become large, and this affects the speed of calculation. Generally we are able to calculate numbers involving smaller digits more easily than those involving larger digits. Recognising this, numbers were often converted to Mishrank (known in modern mathematics as the Redundant Number System) before the calculations were made. Let us illustrate this with an example.

Example : Multiply 484 by 397.

Using Urdhva Tiryaka without Mishrank,

$$\begin{array}{rrrrrr} & & & 4 & 8 & 4 \\ & & & \longleftarrow 7 & 9 & 3 \\ & 7 & 12 & 9 & 2 & \\ \hline 1 & 9 & 2 & 1 & 4 & 8 \end{array}$$

$7 \times 4 = 28$
$7 \times 8 + 9 \times 4 = 92$
$7 \times 4 + 9 \times 8 + 3 \times 4 = 112$
$9 \times 4 + 3 \times 8 = 60$
$3 \times 4 = 12$

Let us now use Mishrank,

$484 = 5\bar{2}4; \quad 397 = 40\bar{3}$

$$\begin{array}{rrrrrr} & & & 5 & \bar{2} & 4 \\ & & & \longleftarrow \bar{3} & 0 & 4 \\ & & & \bar{1} & & \\ \hline 2 & 0 & \bar{8} & 1 & 5 & \bar{2} \end{array} = 192148$$

$\bar{3} \times 4 = \overline{12}$
$\bar{3} \times \bar{2} + 0 \times 4 = 6$
$\bar{3} \times 5 + 0 \times \bar{2} + 4 \times 4 = 1$
$0 \times 5 + 4 \times \bar{2} = \bar{8}$
$4 \times 5 = 20$

Notice that in Urdhva Tiryaka without Mishrank, the intermediate results involved numbers like 92, 112 and 60, whereas with Mishrank the intermediate results rarely even had a carry. Of course for Mishrank to be useful one must competently multiply and add with a combination of positive and negative numbers. Let us take another example.

Example: Multiply 25938 with 2837.

Note that $25938 = 3\bar{4}\bar{1}4\bar{2}$ and $2837 = 3\bar{2}4\bar{3}$

$$\begin{array}{cccccccc} & & & 3 & \bar{4} & \bar{1} & 4 & \bar{2} \\ & & & & & & & \quad \longleftarrow \bar{3}\ 4\ \bar{2}\ 3 \\ \bar{1} & 1 & \bar{1} & & & 2 & \bar{2} & \\ \hline 8 & \bar{7} & 6 & \bar{1} & \bar{4} & 1 & 0 & 6 \end{array} \quad = 73586106$$

Check the result using Navasesh. **N**(25938) = 9 and **N**(2837) =2 and their product yields a Navasesh of 9. Note that **N**(73586106) =9. Since the conversion of numbers to Mishrank form also requires practice, it is often desirable that in the beginning we use Navasesh to check the conversion. Let us see how.

N($3\bar{4}\bar{1}4\bar{2}$) = 3 − 4 − 1 + 4 − 2 = 0
which is equivalent to 9, or **N**(25938).

N($3\bar{2}4\bar{3}$) = 3 − 2 + 4 − 3 = 2
which is same as **N**(2837), and

N($8\bar{7}6\bar{1}\bar{4}106$) = 8 − 7 + 6 − 1 − 4 + 1 + 0 + 6 = 9
which is same as **N**(73586106).

Traditional way of carrying out Urdhva Tiryaka multiplication

As explained, the Urdhva Tiryaka multiplication method involves reversing of one number and shifting it from right to left below the other number. Traditional way of carrying out Urdhva Tiryaka multiplication is however different. Urdhva Tiryaka literally means *vertically and cross-wise*. We illustrate this method of carrying out multiplication using the same example as in the beginning of the chapter. Successive digits of the answer are obtained by multiplying the digits at either end of each double-headed arrow, and adding the products (and carry), as indicated below:

```
5 3 4          4 × 3 = 12
    |
4 6 3
    1
-----
    2

5 3 4          3 × 3+4 × 6 = 33
  X            33 + 1 = 34
4 6 3
3 1
-----
  4 2

5 3 4          5 × 3+3 × 6+ 4×4 = 49
 \|/           49 + 3 = 52
4 6 3
5 3 1
-----
2 4 2

  5 3 4        5 × 6+3 × 4 = 42
   X           42 + 5 = 47
  4 6 3
4 5 3 1
-------
7 2 4 2
```

```
  5 3 4
  |
  4 6 3

  4 5 3 1
-----------
2 4 7 2 4 2
```

$5 \times 4 = 20$
$20 + 4 = 24$

The method can be easily extended to multiplication involving any number of digits. Some may find this method simpler as it does not involve reversing of numbers. Others may find the method of reversing and shifting simpler.

It is important to know this cross-multiplication approach also, since this cross-multiplication technique is used later in the chapter on division.

Exercises

1. Perform the following multiplications using Urdhva Tiryaka and check your results using Navasesh.

   a) 23×42
   b) 314×523
   c) 521×317
   d) 4123×5214
   e) 3147×4253
   f) 42121×21314

2. Convert the following numbers to Mishrank so that each digit is less than 5.

a) 2837

b) 3492

c) 2487

d) 8209

e) 2897

f) 38498

3. Perform the following multiplications using Urdhva Tiryaka and Mishrank and check your results using Navasesh.

a) 3292 × 184

b) 2348 × 3721

c) 3488 × 5019

d) 29302 × 31879

e) 30493 × 24938

f) 310294 × 2401

Chapter 4

Squaring Using Urdhva Tiryaka

The Urdhva Tiryaka multiplication technique described in Chapter 3 can be straightaway used to compute the square of a number, since squaring is multiplication of the number with itself. However, the number of computations required can be nearly halved recognising the symmetry involved in the squaring of a number.

Let us first introduce an operator called *Dwandwa or Duplex*, represented here as **D**(). The Dwandwa of a single-digit number is simply the square of the number.

$$\mathbf{D}(a) = a^2,\ \mathbf{D}(5) = 25,\ \mathbf{D}(7) = 49 \text{ etc.}$$

The Dwandwa of a two-digit number, ab, is

$$\mathbf{D}(ab) = 2 \times a \times b,$$

or $\mathbf{D}(34) = 2 \times 3 \times 4 = 24,$

or $\mathbf{D}(57) = 70.$

The Dwandwa of a three-digit number, abc, is

$$D(abc) = 2 \times a \times c + b^2,$$

or $$D(345) = 2 \times 3 \times 5 + 4^2 = 46,$$

or $$D(427) = 60$$

The Dwandwa of four digit number abcd, is

$$D(abcd) = 2 \times a \times d + 2 \times b \times c$$

or $$D(2147) = 2 \times (2 \times 7 + 1 \times 4) = 36$$

or $$D(7924) = 92$$

The Dwandwa of 5-digit and 6-digit numbers are similarly defined below :

$$D(abcde) = 2 \times (a \times e + b \times d) + c^2$$

$$D(abcdef) = 2 \times (a \times f + b \times e + c \times d)$$

$$D(abcdefg) = 2 \times (a \times g + b \times f + c \times e) + d^2$$

The Dwandwa of a number involving a large number of digits can be determined by taking twice the product of the first and last digits, second and last-but-one digit, third and last-but-two digits, and so on. If a single digit remains in the middle (for numbers with an odd number of digits), the square of this digit is taken. These products are all added to obtain the Dwandwa of the number.

Use of Dwandwa to find the square of a number

The Dwandwa operator can be utilised straightaway to find the square of a number. Let us illustrate this by finding the square of 325. First only one digit is taken from the right and its Dwandwa is taken. The first digit of this Dwandwa is the first digit of the square, whereas the higher digits are taken as carry :

```
 [3 2]5          D(5) = 25
    2
  ---
    5
```

One can imagine that only the rightmost digit of 325 is visible, with a scale covering the remaining digits. The scale is now shifted left revealing one more digit. The Dwandwa of the two visible digits is now found, the carry is added giving us the second digit of the square as well as carry for the third digit :

```
 [3]2 5          D(25) = 20
   2 2
   ---
   2 5
```

The process is repeated with the scale again shifted left. The Dwandwa of the visible digits is found and the carry is added :

```
 [ ]3 2 5        D(325) = 34
   3 2 2
   -----
   6 2 5
```

Once all the digits are revealed, the scale is moved to the right of the number covering one digit. The Dwandwa of the visible digits is found and the carry is added :

```
   3 2[5]        D(32) = 12
 1 3 2 2
 -------
 5 6 2 5
```

The scale is again shifted left, now covering two digits and the process is repeated.

```
     3 2 5 [scale]            D(3) = 9

  1 3 2 2
  -----------
  1 0 5 6 2 5
```

Since further shifting of the scale will cover all the digits, the last step was the final step and the result is 105625. To check the calculation we first obtain the Navasesh of 325 which is 1. The Navasesh of its square is also 1. Since **N**(105625) is also 1, the result is verified.

The procedure is simple, the scale can be replaced by one's hand or can be just visualised, and the squaring operation can be performed almost entirely mentally. Let us consider another example to illustrate the technique.

Example : Find the square of 5432.

```
        5 4 3 2               D(2)=4
                              D(32)=12

   4  5 4 2 1                 D(432)=25
  ---------------             D(5432)=44
  2 9 5 0 6 6 2 4             D(543)=46
                              D(54)=40
                              D(5)=25
```

Navasesh

N(5432)=5, **N**[**N**(5432) × **N**(5432)]=7 and **N**(29506624)=7.

The squaring operation is carried out and verified as above. We will illustrate the power of the method below by finding the square of a large number mentally.

Example: Find the square of 23418564.

$$\begin{array}{l} \qquad\qquad\qquad\qquad 2\ 3\ 4\ 1\ 8\ 5\ 6\ 4 \\ 1\ \ 2\ \ 3\ \ 6\ \ 8\ \ 13\ \ 12\ \ 15\ \ 13\ \ 14\ \ 13\ \ 8\ \ 4\ \ 1 \\ \hline 5\ 4\ 8\ 4\ 2\ \ 9\ \ 1\ \ 3\ \ 9\ \ 8\ \ 2\ 2\ 0\ 9\ 6 \end{array}$$

Use of Mishrank

Once again, the squaring operation involving digits greater than 5 can be simplified by using Mishrank. Let us illustrate this with an example.

Example: Find the square of 379.

First, note that $379 = 4\overline{21}$

$$\begin{array}{l} \qquad\quad 4\ \bar{2}\ \bar{1} \\ \quad \bar{1} \\ \hline 1\ 5\ \bar{6}\ \bar{4}\ 4\ 1 = 143641 \end{array} \qquad \begin{array}{l} D(\bar{1}) = 1 \\ D(\bar{2}\bar{1}) = 4 \\ D(4\bar{2}\bar{1}) = \bar{4} \\ D(4\bar{2}) = \overline{16} \\ D(4) = 16 \end{array}$$

The result can be verified using Navasesh as **N**(379) = 1, **N**[**N**(379) × **N**(379)] = 1 and **N**(143641) = 1. Without Mishrank, the effort involved would be greater.

Let us now find the square of the 8-digit number considered above, but using Mishrank.

Example : Find the square of 23418564.

Note that $23418564 = 2342\overline{144}$

$$\begin{array}{l} \qquad\qquad\qquad 2\ 3\ 4\ 2\ \bar{1}\ \bar{4}\ \bar{4}\ 4 \\ 1\ 2\ 3\ 2\ \bar{1}\ \bar{4}\ \bar{4}\ \bar{2}\ 2\ 4\ 2\ \bar{1}\ \bar{3}\ 1 \\ \hline 5\ 4\ 8\ 4\ 3\ \bar{0}\ \bar{8}\ \bar{6}\ \bar{1}\ 8\ 2\ 3\ \bar{9}\ \bar{1}\ 6 = 548429139822096 \end{array}$$

Note that the result is the same as obtained earlier. If one is well-versed with the handling of both positive and negative numbers, the use of Mishrank makes computations easier, and more interesting.

Exercises

1. Find the squares of the following numbers using Mishrank wherever necessary.

a) 342
b) 2143
c) 30245
d) 214304
e) 382
f) 2738
g) 31948
h) 2031425
i) 2293401
j) 31425028

Chapter 5

Evaluation of Powers

The Urdhva Tiryaka principle used for multiplying and calculating squares in chapters 4 and 5 can also be used to carry out large divisions. This will be thc subject of the next chapter. But before we proceed with divisions, let us illustrate another significant achievement of Indian mathematicians well over 2000 years ago.

The binary number system is the heart of all computers and digital communication systems today. This has been introduced in schools recently and is generally believed to be of 20th century origin. Very few people know that something like the binary representation of a number was used in India to compute the nth power, X^n, of a number X, as illustrated in Pingala's Classic Chandah-Sutra written prior to 200BC.

The procedure is simple. First represent n in binary form. For example, for n=27, its binary representation n=11011.

The binary representation of a number can be obtained by repeatedly dividing the number by 2. The right-most digit (more popularly known as the bit) of the binary number is first obtained by dividing the number by 2. If the remainder is 0, the right-most bit is 0, otherwise it is 1. The quotient is again divided by 2 to obtain the second right-most bit in a similar manner. Like the decimal system, the binary system of numbers uses the place value system and can be written as follows:

$$n=[27]_{10} = 11011 = 1\times2^4+1\times2^3+0\times2^2+1\times2^1+1\times2^0$$

where $[27]_{10}$ indicates that the number is represented in the decimal system.

To compute the nth power of X, n is first represented in the binary number system. The left-most bit which is 1, is deleted. Then each 1 is replaced by SX and each 0 is replaced by S. Therefore for n=27, we obtain,

```
1̸   1    0    1    1
    SX   S    SX   SX
```

The above is the operation that needs to be carried out on X from left to right with Simplying squaring and X implying multiplication by X. Let us illustrate the calculation of X^{27},

	S	X	S	S	X	S	X
X =>	X^2	X^3	X^6	X^{12}	X^{13}	X^{26}	X^{27}

Note that X is first squared, then multiplied by X, then squared twice, then multiplied by X, then squared again, followed finally by multiplication by X, to obtain the desired result.

What is important is that this method of computing X^n requires no other intermediate numbers to be noted (or stored, if the computation is perfoımed by a computer) except for the current partial result. In this sense it is optimum and the fastest method for calculation of X^n. However, if storage of partial results is allowed the calculation can sometimes be carried out with fewer multiplications. Let us illustrate the method with a few examples.

Example: Evalute X^{35}.

35 =	1	0	0	0	1	1
	S X S	S	S	S X	S X	

Operation :	S	S	S	S	X	S	X
X =>	X^2	X^4	X^8	X^{16}	X^{17}	X^{34}	X^{35}

Example : Evaluate X^{49}.

49 =	1	1	0	0	0	1
	S X S	X S	S	S	S	X

Operation :	S	X	S	S	S	S	X
X =>	X^2	X^3	X^6	X^{12}	X^{24}	X^{48}	X^{49}

We now perform the calculations using some small values for X. It is convenient to make a table with the first column being binary representation of the power n. The second column gives the operation to be carried out. The third column gives the power of X obtained at this stage. The fourth column gives the actual calculations. Calculations are performed using Urdhva Tiryaka method. The fifth column gives

Navasesh for verification. The following example illustrates the procedure.

Example: Evaluate 3^{18}.

18 = 1 0 0 1 0

Binary rep. of 18	Operation	Powers of X	Powers of 3	Calculations	Navasesh
1					
0	S	X^2	9		(9)
0	S	X^4	81		(9)
1	S	X^8	6561	81 1 6561	(9)
	X	X^9	19683		(9)
0	S	X^{18}	387420489	1 9 683 2 10 14 20 16 10 4 3 8 7 4 2 0 489	(9)

Note that the Navasesh of 3^2 is 9 and therefore Navasesh of all higher powers of 3 will also be nine as shown in the table above.

Example : Evaluate 4^{12}.

12= 1 1 0 0

Binary rep. of 12	Operation	Powers of X	Powers of 4	Calculations	Navasesh
1					
1	S	X^2	16		(7)
	X	X^3	64		$(7)\times(4)=(1)$
0	S	X^6	4096	6 4 4 1 ——— 4 0 9 6	$(1)\times(1)=(1)$
0	S	X^{12}	16777216	4 0 9 6 7 5 9 11 3 ——————— 1 6 7 7 7 2 1 6	$(1)\times(1)=(1)$

Exercises

1. **Calculate the following using Pingala's method of representing the power in binary form. Use Navasesh to check your calculations.**

    a) 2^{18}
    b) 3^{14}
    c) 4^{11}
    d) 5^{16}
    e) 8^{9}
    f) 24^{7}

Chapter 6

Division Using Urdhva Tiryaka

Division is one of the more difficult arithmetic operations because it involves *estimation* of the quotient at each step. This estimation process is carried out by running through the tables of the divisor and determining the quotient digit which, when multiplied with the divisor gives the largest number smaller than the dividend. For a single-digit divisor, the process is simple because we have memorised the single-digit tables and the operation is virtually a *memory look-up*. As the divisor becomes larger, memory look-up is not possible, and we resort to trial and correction.

The Urdhva Tiryaka principle used earlier for multiplication and squaring can be used to perform division also. The method is called *Straight-Division* and involves division only by a single digit, for which memory recall can be effectively used. This gives us the capability to carry out a large division almost entirely mentally.Let us illustrate this, first with a simple example :

```
⑦2 )  3  3  8  9  4  1  6          33 = 7 × 4 + 5
         5
      ---------------------
         4
```

We wish to divide a number 3389416 by 72. Only the first digit, 7, of the divisor will be used for dividing, while the second digit, 2, will be used just for correcting the dividends. Thus the first two digits of the quotient 33 is first divided by 7 to give a quotient of 4 and remainder of 5. The quotient is written under the second digit below the line, and the remainder is written just below the second digit and above the line as shown above.

The remainder 5 is combined with the next dividend digit 8 to form the partial dividend 58. From this, the product of the previous quotient digit 4 and the second divisor digit 2 (not used so far), is subtracted to obtain the corrected partial dividend, 58 – 4 × 2 = 50, as shown below :

```
            50                      50 = 7 × 7 + 1
⑦2 )  3  3  8  9  4  1  6
           /
         5  1
      ---------------------
         4  7
```

This partial dividend is now divided by 7 to obtain a quotient digit 7 and remainder of 1. This remainder digit and the next dividend digit 9 form the next partial dividend 19, from which the product of the previous quotient digit 7 and the second divisor digit is subtracted to form the corrected partial dividend, 19 – 7 × 2 = 5, as shown below :

```
                  5
⑦2 )  3  3  8  9  4  1  6          05 = 0 × 7 + 5
               /
         5  1  5
      ──────────────────
         4  7  0
```

Next the partial dividend 5 is divided by 7 to give a quotient of 0 and remainder of 5. The remainder digit is combined with the next dividend digit and 0 × 2 is subtracted to obtain the next corrected partial dividend, 54 – 0 ×2 = 54, as shown below :

```
                     54
⑦2 )  3  3  8  9  4  1  6          54 = 7 × 7 + 5
                  /
         5  1  5  5
      ──────────────────
         4  7  0  7
```

Dividing 54 by 7 yields 7 as quotient and 5 as remainder. The remainder is combined with the next dividend digit 1 and 7 × 2 is subtracted to obtain the corrected partial dividend 51 – 7 × 2 = 37. Dividing this by 7 yields 5 as the next quotient digit and 2 as the remainder, which is combined with the next divisor digit 6. From this 5 × 2 is subtracted to obtain the corrected partial dividend 16.

```
                     37 16
⑦2 )  3  3  8  9  4  1  6
                     /  /
         5  1  5  5  2
      ──────────────────
         4  7  0  7  5
```

The division process is now complete as there are no more dividend digits. The quotient is 47075 and the remainder is the corrected partial dividend 16 obtained at this point. In other words, we can write,

	3389416	=	72 × 47075 + 16
Navasesh	(7)	=	(9) × (5) + (7) = (7)

Using Navasesh, one can check that the calculation is indeed correct, since the Navasesh of the RHS is equal to the Navasesh of the LHS.

This straight-division method may appear to be complex at first sight. However, with a little practice, these calculations can be carried out mentally. Let us show this with another example.

Example : Divide 42915 by 83.

```
                 14 58 04
(8)3 )  4  2  9  1  5
              /  /  /
           2  6  2
       ------------------
           5  1  7
```

42 = 8 × 5 + 2
29 − 5 × 3 = 14

14 = 8 × 1 + 6
61 − 1× 3 = 58
58 = 8 × 7 + 2
25 − 7 × 3 = 04

The quotient is therefore 517 and the remainder is 04. Check this using Navasesh as follows :

	42915	=	83 × 517+ 04
Navasesh:	(3)	=	(2) × (4) + (4)

If instead of obtaining the quotient and the remainder, we wish to continue the division and obtain the result upto several decimal places, the process can be continued, noting that an arbitrary number of zeros can be appended to the right of dividend after placing a decimal. This is shown below using the first example once again.

$$
\begin{array}{rrrrrrrrrrrr}
 & & & & & & & 16 & & 16 & 16 & \\
⑦2) & 3 & 3 & 8 & 9 & 4 & 1 & 6 & . & 0 & 0 & 0 \\
 & & & 5 & 1 & 5 & 5 & 2 & & 2 & 2 & 2 \\
\hline
 & & & 4 & 7 & 0 & 7 & 5 & . & 2 & 2 & 2
\end{array}
$$

The quotient is therefore 47075.222.

Problems with negative partial dividends

So far the partial dividends that we obtained, remained positive even after correction. Since the correction of partial dividends involves subtraction, the partial dividend may become negative. In this section, we learn to deal with these negative partial dividends. Let us illustrate the technique using an example we have already seen.

$$
\begin{array}{rrrrrrrr}
 & & & & & 04 & & 40 \\
⑧3) & 4 & 2 & 9 & 1 & 5 & . & 0 \\
 & & 2 & 6 & 2 & 4 & & 0 \\
\hline
 & & 5 & 1 & 7 & . & 0 & 5
\end{array}
$$

The quotient is 517.05. However if we want to continue to divide to obtain the quotient upto further decimal places, we land up with a problem. Since the previous remainder digit is 0 and the next dividend digit is also 0, the current partial dividend is 00. From this we have to subtract the product of the previous quotient digit 5 and the second divisor digit 3, resulting in a corrected partial quotient –15.

```
                     -15
(8)3 )  4  2  9  1  5 . 0  0  0
           2  6  2  4  0
        -----------------------
           5  1  7 . 0  5
```

Since we cannot handle negative partial dividends, we realise that we have over-estimated the previous quotient digit. We go-back one step, reduce this quotient by 1. This gives a remainder of 8 and we continue as follows :

```
                     68 16
(8)3 )  4  2  9  1  5 . 0  0  0
           2  6  2  4  8  6  7
        -----------------------
           5  1  7 . 0  4  8  2
```

The quotient thus obtained is 517.0482 and if necessary, more digits can be obtained. Let us take another example to illustrate the go-back process :

Example : Divide 493372 by 96.

```
           13 37  7̄
(9)6 )  4  9  3  3  7  2
           4  4  1
        -----------------
           5  1  4
```

After three quotient digits we obtain a negative corrected partial dividend written above in Mishrank form and we go back one step to reduce the quotient digit :

			13	37	89	28
(9)6)	4	9	3	3	7	2
		4	4	10	8	
		5	1	3	9	

We notice that going back one step leaves a remainder of 10 and therefore a partial dividend of 107. We subtract 3 × 6 from this to obtain 89 as the corrected partial dividend. The process of straight-division is thereafter continued to obtain a quotient of 5139 and a remainder of 28. Therefore,

$$493372 = 96 \times 5139 + 28$$

$$\text{Navasesh:} \quad (1) = (6) \times (9) + (1) = (1)$$

The calculation is verified using Navasesh.

It should be pointed out here that due to go-back, sometimes the corrected partial dividend may exceed 10 times the first divisor digit. In such cases, the maximum quotient digit is taken as 9, whereas the remainder is allowed to exceed the divisor digit. We will see an example of this later in the chapter.

Note that the go-back process involves repeating the previous division step and recalculation of the next corrected partial dividend. It is possible to avoid this repeatition. Whenever a negative corrected partial dividend is obtained, one can reduce the previous quotient digit by 1 and the new partial dividend can be obtained by adding to it the sum of the following,

(i) first divisor digit multiplied by 10,

(ii) second divisor digit.

Thus in the above example, when the partial product –7 is obtained, the previous quotient digit is reduced by 1 and the new partial dividend is obtained as –7 + 96 or 89 as follows :

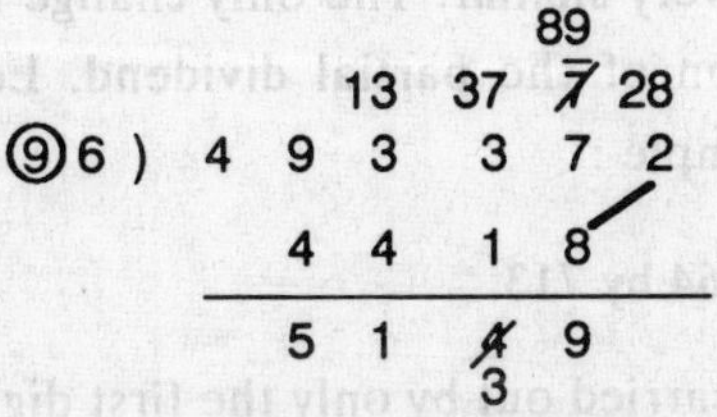

17 – 4×6=–07
–07+96=89

Let us illustrate this procedure of correcting a partial dividend for negative remainders with yet another example :

Example : Divide 22495 by 37.

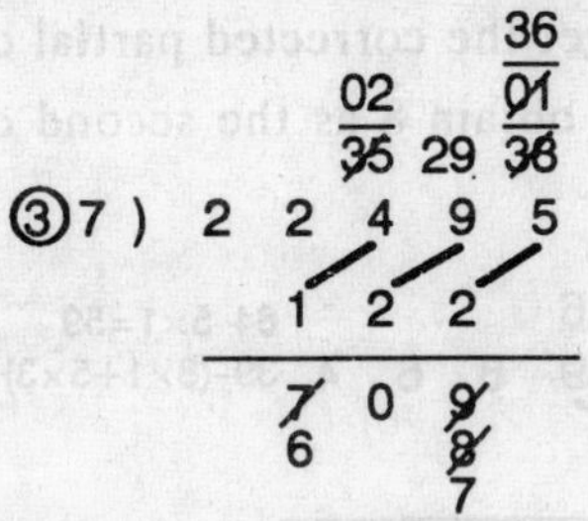

14 – 7 × 7 = – 35
– 35+37=02

29 – 0 × 7 = 29
25 – 9 × 7 = –38
–38 +37 = –1
–1 +37 = 36

The quotient is 607 and the remainder is 36. In other words,

$$22495 = 37 \times 607 + 36$$

Navasesh: (4) = (1) × (4) + (9) =(4)

The division is verified using Navasesh. It may be noted that due to frequent go-backs, the above calculation is difficult. The problem can be simplified by carrying out normalisation as dicussed later in the chapter.

Division by three-digit divisors

So far we have looked at the straight-division method involving only 2-digit divisors. We will now consider 3-digit divisors. The process is very similar. The only change is in the step involving correction of the partial dividend. Let us illustrate this with an example :

Example : Divide 4149864 by 713.

The division is again carried out by only the first digit of the divisor 7. For the first few steps, the calculation is similar to that used in the case of two-digit divisor. The partial dividend 41 is divided by 7 to get 5 as the first quotient and 6 as the remainder digit. The next partial dividend 64 is corrected by subtracting the product of the previous quotient digit 5 and the second divisor digit 1 to get the corrected partial dividend of 59. This is divided by 7 to obtain 8 as the second quotient digit and 3 as the remainder.

```
                 59 16            64-5×1=59
⑦1 3 )   4  1  4  9  8  6  4    39-(8×1+5×3)=16
             6  3
          ---------------------
             5  8
```

The next partial dividend is 39 which is to be corrected, this time by subtracting the *cross-product of the previous two quotient digits and the last two digits of the divisor*. This correction is new. It involves multiplication of the last quotient-digit, 8, with the second divisor-digit, 1, and the last-but-one quotient-digit, 5, with the third divisor-digit, 3. The two products are subtracted from the partial dividend to obtain

39 – (8×1+5×3) or 16. The division process is then continued as before except that the correction in partial dividends at each step will involve two products as described above. The continuation of the division is shown below :

```
                 16 02 20                  28–(2×1+8×3)=2
⑦1 3 )   4  1  4  9  8  6  4              26–(0×1+2×3)=20
            6  3  2  2
         --------------------
            5  8  2  0
```

The next quotient digits is 2 followed by a quotient digit of 0. The corrected partial dividend is now 20.

We have already used all but one dividend digit. Since the divisor has three digits, the quotient will not have any more digits, unless we wish to obtain digits after the decimal point. The computation of remainder is however not completed. Since no division is to be carried out, the corrected partial dividend 20 obtained at this stage, appended with the next digit 4, is the uncorrected remainder. This is corrected in the same way in which the partial dividend is corrected. Imagining that 0 is the new quotient-digit, the last two quotient-digits are cross-multiplied with the last two digits of the divisor and subtracted from the remainder to obtain, 204 – (0 × 1+0 × 3)=204.

Thus 5820 is the quotient and 204 is the remainder. In other words,

$$4149864 = 713 \times 5820 + 204$$

Navasesh: $(9) = (2) \times (6) + (6) = (9)$

The division is checked above using Navasesh.

Let us take another example where we illustrate division by a 3-digit divisor, as well as show the go-back process.

Example: Divide 32584945 by 724.

```
                           62
              37 02 04 49  1̶0̶
 ⑦2 4 )    3  2  5  8  4  9  4  5
               4  2  2  4  0
           ------------------------
               4  5  0  0  7̶
                           6
```

Remainder: $625 - (0 \times 2 + 6 \times 4) = 601$

Note that the first 4 digits of the quotient are obtained in a straight forward manner. The next corrected partial dividend is 49 and dividing this by 7 results in a quotient digit of 7 and 0 remainder. The next corrected partial dividend works out to be –10. To make it positive, the last quotient digit is reduced by 1 to 6. The corrected partial dividend is now –10 plus the first two digits of the divisor, $-10+72=62$.

Since we have already reached the last-but-one dividend-digit and the divisor is a 3-digit number, we stop the division process. The uncorrected remainder is 625. The corrected remainder is obtained by subtracting the sum of the cross-product of the last two quotient digits, with a zero being imagined as the last quotient digit and the last two divisor digits. The quotient is therefore 45006 and the remainder is 601 and we can write,

$$32584945 = 724 \times 45006 + 601$$

Navasesh: $(4) = (4) \times (6) + (7) = (4)$

The check is carried out using Navasesh.

Let us take a still more complicated example. In this example we will get a negative dividend when the previous quotient digit is 0. This would involve going back two steps or correction for two quotient digits.

Example: Divide 3257473 by 724.

$$\begin{array}{lccccccc} & & & & 37 & 01 & \overline{06} & \\ ⑦24\,) & 3 & 2 & 5 & 7 & 4 & 7 & 3 \\ & & 4 & 2 & 1 & & & \\ \hline & & 4 & 5 & 0 & & & \end{array}$$

Note that the first three quotient digits can be obtained in a straight forward manner. The next corrected partial dividend is now negative, i.e., $\overline{06}$. We wish to reduce the previous quotient digit, this however is now zero. Therefore, we have to move towards the left and reduce the right-most non-zero quotient digit by 1. We then go back to this step and continue the calculation as shown below :

$$\begin{array}{lccccccc} & & & & 37 & 73 & 70 & 23 \\ ⑦24\,) & 3 & 2 & 5 & 7 & 4 & 7 & 3 \\ & & 4 & 9 & 10 & 7 & & \\ \hline & & 4 & 4 & 9 & 9 & & \end{array}$$

Remainder=233 – (0 × 2 + 9 × 4) = 197.

Note that after the second quotient-digit was calculated, the corrected partial dividend was 73. Dividing by 7 would now result in a quotient of 10. But in the straight-division method, *the maximum quotient digit is taken as 9* whereas the remainder digit may exceed the divisor 7, or even 10. Continuing this, we obtain a quotient of 4499 and a remainder of 197, and we can write

$$3257473 = 724 \times 4499 + 197$$

Navasesh: (4) = (4) × (8) + (8) = (4)

Instead of going back, we could have continued the operation by making some corrections. The correction involved in the partial dividend is however different when two quotient digits are corrected. It involves the addition of the following to the negative partial dividend,

(i) the first divisor-digit multiplied by 10,

(ii) the second divisor digit,

(iii) the third divisor digit.

Thus the corrected partial dividend when we get the negative partial dividend, –06, will be,

$$-06 + 7 \times 10 + 2 + 4 = 70$$

and we can continue the division as shown :

```
                         70
                         ‾‾
                        0̸6 23
⑦2 4 )   3  2  5  7  4  7  3
             4  2  1  7/
        ───────────────────────
             4  5̸  0̸  9
                4  9
```

Remainder = 233 – 9 × 4 = 197. The result is the same as that obtained earlier.

Let us take yet another example involving multiple go-back steps,

Example: Divide 345274 by 428.

28   −56+42= −14
~~14~~ 35   −14+42=28
09 ~~56~~ ~~07~~   −07+42=35
④2 8)   3 4 5 2 7 4
2 1 0
8 ~~2~~ ~~7~~
~~1~~ 6
0

Remainder=354 − (0 × 2 + 6 × 8) = 306.

Therefore,

$$345274 = 428 \times 806 + 306$$

Navasesh: (7) = (5) × (5) + (9) = (7)

Note that the second quotient digit was first obtained as 2 and the next corrected partial dividend as −56. We reduce the previous quotient to 1, but the corrected partial dividend is still negative or −14. We again reduce the previous quotient digit to 0 and now get the corrected partial dividend as 28. Dividing this by 4 gives the next quotient digit as 7 and the next partial dividend as −07. We again reduce the previous quotient digit to 6 and get the corrected partial dividend as 35. The division here is therefore somewhat complicated. It could be simplified by either use of Mishrank described later in the chapter or by the process of normalisation.

If for any of these divisions we wish to find the quotient digit after the decimal point, instead of the remainder, we could continue the division after placing a decimal followed by zeros in the dividend as shown below :

```
                                  6
                                 1̸0 25 02 08 08
⑦2 4 )  3  2  5  8  4  9  4   5 . 0  0  0
           4  2  2  4  0  6  4   2  1  1
        ---------------------------------
           4  5  0  0  7̸ . 8  3  0  1  1
                       6
```

The answer is 45006.83011.

Division by divisors with more than three digits

The straight-division algorithm can be extended for divisors with any number of digits. The division is always carried out with the first divisor digit. The partial dividends are however corrected by cross-multiplying the (n–1) previous quotient digits with the rightmost (n–1) divisor digits, where n is the number of divisor digits. The only problem is that going back will occur more frequently as the correction involves the sum a of larger number of products. Also, for an n-digit divisor, we need to stop the division when we reach the last (n–1) dividend digits to obtain the remainder. Of course, if fractional digits of the quotient are desired, then the division process is simply continued. Let us illustrate this with several examples.

Example: Divide 5321744 by 8234.

```
                      55                          52–6×2=40
                  40  2̸7̸ 29                      01–5×2–6×8=–27
⑧2 3 4 )  5  3  2   1  7  4  4                  –27+82=55
             5  0   7                             77–6×2–4×3–6×2
        ------------------------                     =29
             6  5̸   6
                4
```

The Remainder is equal to

$$2944-(0\times2+6\times3+4\times4)\times10-(0\times2+0\times3+6\times4)=2580$$

Therefore,

$$5321744 = 8234 \times 646 + 2580$$

Navasesh: $(8) = (8) \times (7) + (6)$

Note that the remainder calculation is slightly complex. The method involves imagining that the division process is continued with zero-valued quotient digits. Thus after obtaining three quotient digits you obtain the partial dividend of 29. Since the quotient digit is assumed to be zero, the remainder is 29. Appending the next quotient digit 4, we obtain the uncorrected partial dividend 294. We subtract from this 0×2+6×3+4×4, since 0 is assumed to be the last quotient digit. The corrected partial dividend is now 294 – 34 or 260. Again the quotient is assumed to be zero and the remainder is 260. Appending the next dividend digit we get the uncorrected dividend of 2604. The correction involves subtracting 0×2+0×3+6×4, since the previous two quotient digits are assumed to be 0. The corrected remainder is therefore 2604 – 24 or 2580.

Example: Divide 3927464 by 72511 correct to four decimal places.

```
                        47     53
                 32 14  2̸5 30  1̸9
(7)2 5 1 1 )   3  9  2   7  4   6  4
                  4  4   0  5   2  4
               ---------------------
                  5  4̸ . 2  6   4̸  7
                         1      3
```

42 – 5×2=32
47 – 4×2 – 5×5=14
04 – 2×2 – 4×5 – 5×1=–25
56 – 6×2 – 1×5 – 4×1 – 5×1=30
24 – 4×2 – 6×5 – 1×1 – 4×1=–19

The answer is 54.1637...

Note that the last digit of the quotient may need correction if the division is continued.

Straight-division using Mishrank

The straight-division technique allows us to carry out large divisions mentally. What slows down the division are go-backs. Frequent go-backs can be avoided if one is willing to use Mishrank. In this case, one does not necessarily have to go back when a negative corrected partial dividend is encountered. One can instead use a negative quotient and proceed. Let us use this to carry out the division in the previous example.

		32	14	$\overline{25}$	25	52	03		
(7)2 5 1 1)	3	9	2	7	4	6	4	.0	0
			4	4	0	3	4	3	3
		5	4	.2	$\overline{4}$	3	7	0	

04 – 2×2 – 4×5 – 5×1= –25
36 – (–4)×2 – 2×5 – 4×1 – 5×1=25
44 – 3×2 – $\overline{4}$×5 – 2×1 – 4×1=5
30 – 7×2 – 3×5 – $\overline{4}$×1 – 2×1=03

Note that the fourth partial dividend is –25. Instead of going back we use $\overline{4}$ as quotient and proceed as before. Note that the partial dividend correction will also now involve negative numbers. The quotient is obtained as $54.2\bar{4}370$ or 54.1637 which is the same as that obtained earlier.

Example: Divide 3419753 by 647.

$$
\begin{array}{r|ccccccccccc}
 & & & 21 & \bar{8} & 34 & 39 & \overline{26} & 18 & 23 & 17 \\
\textcircled{6}47\ ) & 3 & 4 & 1 & 9 & 7 & 5 & 3. & 0 & 0 & 0 \\
 & & 4 & 3 & 4 & 4 & 3 & 4 & 0 & 5 & 5 \\
\hline
 & & 5 & 3 & \bar{2} & 5. & 6 & \bar{5} & 3 & 3 & 2
\end{array}
$$

The quotient is $53\bar{2}5.6\bar{5}33$ or 5285.5533.

One can also carry out the above division by writing the divisor 647 as $65\bar{3}$

$$
\begin{array}{r|cccccccccc}
 & & & 16 & 54 & \overline{32} & 37 & \overline{32} & 23 & 20 \\
\textcircled{6}5\,\bar{3}\ ) & 3 & 4 & 1 & 9 & 7 & 5 & 3. & 0 & 0 \\
 & & 4 & 4 & 0 & \bar{2} & 1 & \bar{2} & 5 & 2 \\
\hline
 & & 5 & 2 & 9 & \bar{5}. & 6 & \bar{5} & 3 & 3
\end{array}
$$

The quotient is 5285.5533 which is same as before.

Note that even while using Mishrank, one may need to go back in certain cases. This occurs when the previous quotient digit has been over-estimated or under-estimated to an extent that Mishrank correction is not sufficient. The problem of when to go back is a bit more complex. As a rule, if the corrected partial dividend is negative and its magnitude is more than ten times the divisor digit, one should go back and correct.

Also note that in the previous division when we divided $\overline{26}$ by 6, we wrote a quotient of $\bar{5}$ and a positive remainder of 4, whereas in the above division while dividing $\overline{32}$ by 6, we wrote the quotient as $\overline{5}$ and the remaider as $\overline{2}$. The choice made in these cases were to a certain extent, based on what to expect as we proceed in the division, and comes with experience. Poor choices will lead to more frequent go-backs.

Normalisation of divisions

The difficulties described above occur more frequently when the lead divisor digit is small, say 1, 2, 3, or 4. In such cases, it may be preferable to normalise the divisor and the dividend. Both the divisor and the dividend may be multiplied by a small number (like 2 or 3) such that the lead divisor digit becomes 5 or greater. This would, however, require correction in the remainder (denormalisation) once the division is completed. Let us show this with an example :

Example : Divide 15429 by 26.

Let us first try the straight-division without normalising.

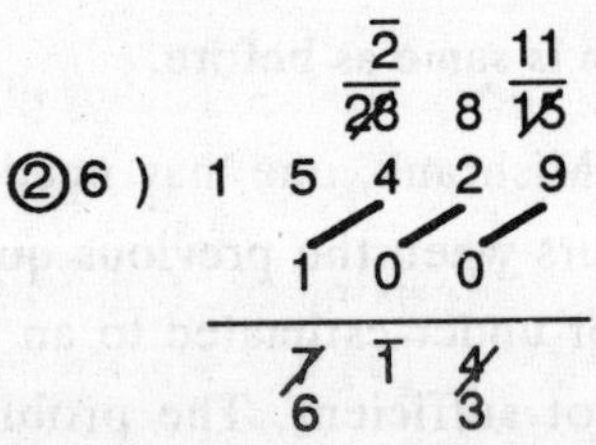

The quotient is $6\overline{1}3$ or 593 and the remainder is 11. Note that at one stage the corrected dividend was $\overline{28}$ which is larger than 2 × 10. We go back one step and correct the quotient.

Now let us normalise by multiplying both the dividend and the divisor by 2. The new divisor is 26 × 2 or 52 and the new dividend is 15429 × 2 or 30858. Let us try the straight-division with these numbers :

$$\begin{array}{lccccc} & & & \bar{4} & 17 & 22 \\ \textcircled{5}2\ ) & 3 & 0 & 8 & 5 & 8 \\ & & 0 & 1 & 2 & \\ \hline & & 6 & \bar{1} & 3 & \end{array}$$

The quotient is 6$\bar{1}$3 or 593 and the remainder is 22. The remainder needs to be denormalised by dividing by 2. The corrected remainder is same as what we found earlier, i.e., 11.

We can also multiply the divisor and the dividend, in this case, by 3 to get a divisor of 78 and a dividend of 46287. The division process will be as follows :

$$\begin{array}{lccccc} & & & \bar{6} & 26 & 33 \\ \textcircled{7}8\ ) & 4 & 6 & 2 & 8 & 7 \\ & & 4 & 1 & 5 & \\ \hline & & 6 & \bar{1} & 3 & \end{array}$$

The quotient is 6$\bar{1}$3 or 593 and the corrected remainder is 33 divided by 3 or 11.

Note that if we do not wish to determine the remainder, but only the quotient up to certain decimal places, the division can be carried out after normalisation without any correction at the end.

The straight-division method is a powerful one and has been used in India for a long time. It has recently been used to carry out large divisions on computers and is proving to be superior to other known division algorithms.

Exercises

1. Use straight-division to carry out the following divisions and obtain the quotients and remainders. Check your result using Navasesh.

    a) 5923 / 84
    b) 71352 / 96
    c) 41352 / 73
    d) 349253 / 62
    e) 31745 / 44

2. Use straight-division to carry out the following divisions after appropriate normalising and obtain the quotients and remainders. Check your result using Navasesh.

    a) 22439 / 36
    b) 35295 / 43
    c) 15322 / 24
    d) 11534 / 19

3. Use straight-division (and go-back wherever necessary) to obtain the quotients and remainders for the following divisions. Check your result by using Navasesh.

    a) 31429 / 78
    b) 51492 / 66
    c) 35924 / 534
    d) 425934 / 851
    e) 314902 / 942
    f) 635492 / 8413

4. Redo the divisions in question 3, with on the spot correction of negative partial dividends.

5. Redo the divisions in question 3, using Mishrank wherever negative partial dividends are encountered.

6) Use straight-division (with normalisation and Mishrank wherever necessary) to carry out the following divisions and obtain the quotients upto three decimal places.

   a) 327459 / 87
   b) 419235 / 57
   c) 243924 / 37
   d) 144925 / 282
   e) 529431 / 474
   f) 314251 / 382
   g) 429345 / 7201
   h) 325946 / 8142

Chapter 7

Square Root

A method analogous to the Straight Division algorithm based on Urdhva Tiryaka can be used to find the square root of a number. This is a powerful algorithm and enables us to obtain the square root of a number up to the desired decimal place much faster than most of the other methods. The calculation proceeds very much like in straight division.

Let us find the square root of 70. Noting that there are only two digits before the decimal, we start by finding the largest digit whose square is less than 70. It is obviously 8 with a remainder of 6, i.e. $70 - 8^2 = 6$. Thus 8 is the first digit (the digit before the decimal) of the square root. We use 2 times 8, or 16, as the divisor for the rest of the algorithm as shown below :

				111	114	$\overline{58}$
16)	7	0	.0	0	0	0
		6	12	15	2	6
		8	3	6	7	$\overline{4}$

120 – D(3) = 111
150 – D(36) = 114
20 – D(367) = – 58

As mentioned above, 6 is the remainder; the next dividend digit, i.e., 0, is appended to it to form the partial dividend, 60. We divide 60 by 16 to obtain 3 as the quotient and 12 as remainder. Once again 120 is the partial dividend. However, as in division this needs to be corrected. The correction in the case of square root, involves subtraction of the Dwandwa of the quotient (square root) digits obtained so far leaving out the lead quotient digit (see Chapter 4 for an explanation of Dwandwa). Since the Dwanda of 3 is 9, the corrected partial dividend is 111 and dividing this by 16 gives 6 as quotient and 15 as the remainder.

The next partial dividend is 150. From this the Dwandwa of all but the lead quotient digit, (i.e., 36) is subtracted to obtain the corrected partial dividend of 114. Dividing again by 16 gives 7 as quotient and 2 as remainder. Subtracting the Dwandwa of 367 from the partial dividend 20 gives the corrected partial dividend of –58. Since the partial dividend is negative, we can either go back one step and reduce the quotient (square root) digits, or use Mishrank to obtain the negative quotient digit. Let us do the latter. The quotient digit will be –4 and the remainder will be 6. We have thus obtained the square root upto 4 decimal places as $8.367\bar{4}$ or 8.3666.

It is however quite possible that some of these quotient digits may require correction if the division is continued for a few more decimal places. It can be shown that the error in the quotient at this stage is utmost 0.001, i.e., if the quotient has been calculated upto n decimal places, the error is utmost $10^{(n-1)}$ (for values of n less than 10). Below we show the evaluation of square root upto a few more decimal places, using Mishrank as well as the go-back procedure.

$$
\begin{array}{l}
\qquad\qquad\qquad\qquad\qquad\quad \overline{58}\ \ 0\ \ \bar{1}\ \ 46 \\
16)\ 7\ \ 0.0\ \ 0\ \ 0\ \ 0\ \ 0\ \ 0\ \ 0 \\
\qquad\quad 6\ 12\ 15\ \ 2\ \ 6\ \ 0\ \ \bar{1}\ \ \bar{2} \\
\hline
\qquad\quad 8\,.\,3\ \ 6\ \ 7\ \ \bar{4}\ \ 0\ \ 0\ \ 3
\end{array}
$$

60 − D(367$\bar{4}$) = 60 − 60 = 0

00 − D(36740) = 0 − 1 = $\bar{1}$

$\overline{10}$ − D(367400) = $\overline{10}$ − (−56) = 46

Therefore square root of 70 is 8.3666003. The square root can also be calculated without Mishrank as follows :

$$
\begin{array}{l}
\qquad\qquad\qquad\qquad\qquad\quad 108 \\
\qquad\qquad\qquad\qquad\qquad\quad \overline{58}\ \ 12\ \ 12\ \ 48 \\
16)\ 7\ \ 0.\ 0\ \ 0\ \ 0\ \ 0\ \ 0\ \ 0\ \ 0 \\
\qquad\quad 6\ 12\ 15\ \ 2\ \ 12\ \ 12\ \ 12\ \ 0 \\
\hline
\qquad\quad 8\,.\,3\ \ 6\ \ \underset{6}{\not{7}}\ \ 6\ \ 0\ \ 0\ \ 3
\end{array}
$$

−58 + 160 + 6 = 108

120 − D(3666) = 12

120 − D(36660) = 120 − 108 = 12

120 − D(366600) = 120 − 72 = 48

The square root of 70 is again seen to be 8.3666003. Note that a correction of the partial dividend is required as we go-back one step. The correction involves the addition of the following quantities to the negative partial dividend,

(i) Divisor times 10 or 16 × 10=160

(ii) 2 times the second digit of the square root or 2 × 3=6.

Thus, at the 4th decimal place, we get a corrected partial dividend of −58 +160 +6 or 108.

We further illustrate the Urdhva Tiryaka based square root technique by taking the square root of 54.364.

$$
\begin{array}{lllllllll}
 & & & 107 & 52 & 33 & \bar{4} & 69 & 88 \\
14) & 5 & 4\,.\,3 & 6 & 4 & 0 & 0 & 0 & 0 \\
 & 5 & 11 & 9 & 10 & 5 & 10 & 11 & 4 \\
\hline
 & 7\,. & 3 & 7 & 3 & 2 & \bar{1} & 4 & 6
\end{array}
$$

116 – D(3) = 107
94 – D(37) = 52
100 – D(373) = 33
50 – D(3732) = $\bar{4}$
100 – D(3732$\bar{1}$) = 69
110 – D(3732$\bar{1}$4) = 88

The square root of 54.364 is 7.3731946

Now, let us assume that the number whose square root we wish to determine has three or more digits before the decimal. Group the digits before the decimal into pairs starting from the decimal. First take the square root of the left-most pair (or left-most single digit if the number of digits before the decimal is odd) and proceed as before. The quotient will have as many digits before the decimal as the number of pairs in the original number counting an odd digit as a pair. Let us illustrate the same with the following two examples.

Example : Find the square root of 5489.32.

$$
\begin{array}{llllllllllll}
 & & & & 13 & 133 & 0 & 0 & \overline{81} & 78 & 40 \\
14) & 5 & 4 & 8 & 9\,.\,3 & 2 & 0 & 0 & 0 & 0 \\
 & & 5 & 2 & 13 & 7 & 0 & 0 & 3 & 8 & 12 \\
\hline
 & & 7 & 4\,. & 0 & 9 & 0 & 0 & \bar{6} & 5 & 2
\end{array}
$$

$$
\begin{aligned}
&29 - \mathbf{D}(4) = 13\\
&133 - \mathbf{D}(40) = 133\\
&72 - \mathbf{D}(409) = 72 - 72 = 0\\
&0 - \mathbf{D}(4090) = 0\\
&0 - \mathbf{D}(40900) = -81\\
&30 - \mathbf{D}(40900\bar{6}) = 78\\
&80 - \mathbf{D}(40900\bar{6}5) = 40
\end{aligned}
$$

The square root is therefore 74.0899452.

Note that since there are two pairs of digits before the decimal, the square root has two digits before the decimal.

Example: Find the square root of 948.27.

$$
\begin{array}{rcccccccc}
 & & & & 02 & \overline{37} & \overline{10} & 56 & 36\\
6\,) & 9 & 4 & 8 & 2 & 7 & 0 & 0 & 0\\
 & 0 & 4 & 0 & 2 & \bar{1} & \bar{4} & 2 & 0\\
\hline
 & 3 & 0. & 8 & 0 & \bar{6} & \bar{1} & 9 & 6
\end{array}
$$

$$
\begin{aligned}
&48 - \mathbf{D}(0) = 48\\
&02 - \mathbf{D}(08) = 02\\
&27 - \mathbf{D}(080) = \overline{37}\\
&\overline{10} - \mathbf{D}(080\bar{6}) = \overline{10}\\
&\overline{40} - \mathbf{D}(080\bar{6}\bar{1}) = 56\\
&20 - \mathbf{D}(080\bar{6}\bar{1}9) = 36
\end{aligned}
$$

The square root is 30.793996

Note that since there are two pairs of digits (3 digits) before the decimal the square root has two digits before the decimal. Also, as mentioned earlier, the square root obtained so far may have an error. The error in the quotient now would be less than 0.00001.

Normalisation to determine square roots

When the lead digit in the square root result is 1 or 2, the divisor in this technique will be 2 or 4 respectively. As discussed in the previous chapter, small divisors result in frequent go-backs. It is preferable to normalise the number whose square root we wish to find out in such cases. If the first two digits of the square root have a value less than 25, we can multiply the number whose square root we are seeking by 4. The square root now obtained will be 2 times the square root that we are seeking. Let us show this with an example.

Example: Find the square root of 1.3492.

Since the digit before the decimal is 1, the square root will be 1.xxx. We multiply the number by 4 to obtain 5.3968. Now we find its square root,

			10	14	6	2	7	18
4)	5.	3	9	6	8	0	0	0
	1	1	2	2	2	2	3	
	2.	3	2	3	1	0	1	4

The desired square root is half of 2.3231014 or 1.1615507.

To conclude, the square root technique is fast and relatively simple. With a little practice one should be able to calculate the square root almost mentally.

Exercises

1. Find the square root of the following numbers upto 5 decimal places.

   a) 18.4294
   b) 42.3154
   c) 69.8425
   d) 1634.489
   e) 984.139
   f) 483142.56

2. Find the square root of the following numbers upto 5 decimal places (using normalisation wherever necessary).

   a) 2.3542
   b) 5.4136
   c) 425.312
   d) 7234.48
   e) 57354.26
   f) 78240.42

Chapter 8

Divisibility

Students of mathematics are often concerned with whether one number is divisible by another or not. Fortunately, very simple techniques exist to determine at a glance whether a number is divisible by the numbers 2, 3, 4, 5, 6, 8, 9, 10 etc. However, there are a few prime divisors like 7, 13, 17 etc. for which the divisibility cannot be obtained in a simple manner. This chapter illustrates a technique used in India for quite some time and based on the use of what is known as *Ekadhika* to determine such divisibility. But, before we proceed, let us summarise the well-known techniques for divisibility by numbers like 2, 3, 4 etc.

1) A number is divisible by 2 if its last digit is divisible by 2,

2) A number is divisible by 3 if its Navasesh is divisible by 3 (i.e., either 3, 6 or 9),

3) A number is divisible by 4 if its last two digits are divisible by 4,

4) A number is divisible by 5 if its last digit is divisible by 5 (last digit is either 5 or 0),

5) A number is divisible by 6 if it is divisible by 2 and 3,

6) A number is divisible by 8 if its last three digits are divisible by 8,

7) A number is divisible by 9 if its Navasesh is 9,

8) A number is divisible by 10 if its last digit is 0,

9) A number is divisible by 11 if the difference between the sum of its even digits and the sum of its odd digits is divisible by 11.

We need to further concern ourselves with only 7 and prime numbers greater than 10, as divisibility by other numbers can be straight-away obtained by finding the divisibility by its factors.

To find the divisibility by prime numbers, we first determine their Ekadhikas, represented here by P, and determined as follows :

1) Multiply the divisor by the smallest possible number such that the last digit of the product is 9.

2) Take the higher digits of the product (leaving out the units digits) and add one to it. The number so obtained is the Ekadhika.

Thus for determining the Ekadhika of 7, we multiply it by 7 to obtain 49. The Ekadhika of 7 is therefore 5. Similarly to determine the Ekadhika of 13, we multiply by 3 to obtain 39. Its Ekadhika is 4. The table given below shows the Ekadhikas of

some of the prime numbers. Note that to determine the Ekadhika of the numbers ending in digits 1, 3 , 7 and 9, the multipliers turn out to be 9, 3, 7 and 1 respectively.

Prime number X	Multiplied by	Product	Ekadhika P	Negative Ekadhika Q
7	7	49	5	2
11	9	99	10	1
13	3	39	4	9
17	7	119	12	5
19	1	19	2	17
23	3	69	7	16
29	1	29	3	26
31	9	279	28	3
37	7	259	26	11

Another quantity which is of interest to us is the negative Ekadhika, represented by Q. The Q of a number X, can be obtained as X – P. The Q of some of the prime numbers is also given in the table above.

Let us now look at how the Ekadhika of a number can be used to find the divisibility of other numbers by that number. The procedure is called *Vetsana* or *Osculation*. Let us find out whether a number 4578 is divisible by 7. We find that the

Ekadhika of 7 is 5. We will multiply the units digit of the number 4578 by 5 and add the product to the remaining digits as shown below :

```
4 5 7 8          5 × 8 = 40
    4 0
-------
4 9 7
```

We obtain the three-digit result 497. The process of Vetsana therefore enables us to reduce a n digit number to n–1 digits.

The claim is that if 4578 is divisible by 7, then so is 497. We repeat the process by multiplying the unit digit of 497 by 5 and adding the product to the remaining digits as follows :

```
4 9 7
3 5
-----
8 4
```

We thus obtained the two-digit number 84. It is easy to check that this two-digit number is divisible by 7. This implies that 4578 is also divisible by 7.

Let us take a larger number 4612594 and determine its divisibility by 7. We will once again multiply by 5 from the right and in each step reduce one digit in the quantity whose divisibility is being checked. We do it as follows :

```
Digit no.   7 6 5 4 3 2 1     4 × 5 = 20
            -------------     20 + 9 = 29
Carry               2
            4 6 1 2 5 9 4
                      9
```

At the first step we multiply 4 by 5 and add the next digit to obtain 29. The digit 9 is written below the second digit and the carry 2 is written above the third digit. We have thus reduced the original number by one digit. We once again multiply the right-most digit 9 so obtained, by 5 and add it to the previous digit 5 (third digit), also adding the 2 carried over from the previous step to get 52. We write 2 below the third digit and the carry 5 above the 4th digit as shown below :

Digit No.	7	6	5	4	3	2	1	$9 \times 5 = 45$
								$45 + 5 + 2 = 52$
Carry				5	2			
	4	6	1	2	5	9	4	
					2	9		

We now multiply the right-most digit 2 by 5 and add the 4th digit 2 as well as the carry 2 to obtain 17. We continue this till we get a 2-digit result as shown below :

Digit no.	7	6	5	4	3	2	1
Carry	4	3	1	5	2		
	4	6	1	2	5	9	4
		4	7	7	2	9	
	8	4					

The two digit number thus obtained is 84. We find that this is divisible by 7 and therefore surmise that the original number 4612594 is divisible by 7. We take another example.

Example: Determine whether 3294643 is divisible by 7.

The Ekadhika of 7 is 5.

$$
\begin{array}{ccccccc}
3 & 5 & 1 & 5 & 1 & & \\
3 & 2 & 9 & 4 & 6 & 4 & 3 \\
 & 2 & 5 & 9 & 2 & 9 & \\
\hline
6 & 2 & & & & &
\end{array}
$$

The quantity 62 is not divisible by 7 and therefore the answer to the above question is negative.

Instead of using the positive Ekadhika of 7 for multiplication, we could use the negative Ekadhika. The negative Ekadhika of 7 is 2. The use of negative Ekadhika involves a process just like the above to reduce the number of digits whose divisibility is to be checked. The only difference is that we multiply the right- most digit by the negative Ekadhika 2, and *subtract* it from the other digits, instead of adding to it. The use of Mishrank will obviously help. We illustrate this technique by checking whether 4612594 is divisible by 7.

$$
\begin{array}{ccccccc}
\bar{1} & & & & & & \\
4 & 6 & 1 & 2 & 5 & 9 & 4 \\
 & \bar{2} & 9 & \bar{4} & 3 & 1 & \\
\hline
3 & \bar{2} & & = & 28 & &
\end{array}
\qquad
\begin{array}{l}
9 - 2 \times 4 = 1 \\
5 - 2 \times 1 = 3 \\
2 - 2 \times 3 = \bar{4} \\
1 - 2 \times 4 = 9 \\
6 - 2 \times 9 = \overline{12}
\end{array}
$$

We multiply the right-most digit 4 by 2 and subtract it from previous digit to obtain 1. We next multiply 1 by 2 and subtract it from 5 to get 3. We next subtract 3 × 2 or 6 from the previous digit 2 to obtain $\bar{4}$. We now subtract $\bar{4}$ × 2 or $\bar{8}$ from 1 to obtain 9. Next, we subtract 9 × 2 or 18 from 6 to obtain $\overline{12}$. We write $\bar{2}$ below the next digit and $\bar{1}$ as carry. The original number is therefore reduced to $3\bar{2}$ or 28. Since 28 is divisible by 7, the original number is indeed divisible by 7.

Let us now take examples for ascertaining divisibility by other numbers.

Example: Check whether 106216 is divisible by 11.

The Ekadhika of 11 is 10 and its negative Ekadhika is 1. It is preferable to use negative Ekadhika.

$$\begin{array}{cccccc} 1 & 0 & 6 & 2 & 1 & 6 \\ & 1 & \bar{1} & 7 & \bar{5} & \\ \hline 1 & 1 & & & & \end{array}$$

Since 11 is divisible by 11, 106216 is also divisible by 11. Note that this method for checking divisibility by 11, is equivalent to the simpler method given at the beginning of this chapter.

Example: Check whether 849966 is divisible by 13.

The Ekadhika of 13 is 4 and its negative Ekadhika is 9. It would be preferable to use the positive Ekadhika.

$$\begin{array}{cccccc} {}^{3} & {}^{1} & {}^{1} & {}^{3} & & \\ 8 & 4 & 9 & 9 & 6 & 6 \\ & 7 & 8 & 2 & 0 & \\ \hline 11 & 7 & & & & \end{array}$$

We note that 117 is divisible by 13 and therefore the answer to the question is yes.

Example: Check whether 3492123 is divisible by 13.

We again use the positive Ekadhika of 13 which is 4.

$$
\begin{array}{ccccccc}
1 & 3 & 3 & 1 & 1 & & \\
3 & 4 & 9 & 2 & 1 & 2 & 3 \\
 & 5 & 2 & 5 & 8 & 4 & \\
\hline
4 & 5 & & & & &
\end{array}
$$

Since 45 is not divisible by 13, the answer to the question is no.

Example : Check whether 2394125 is divisible by 17.

The Ekadhika of 17 is 12 and its negative Ekadhika is 5. It is preferable to use the latter.

$$
\begin{array}{ccccccc}
 & 3 & \bar{1} & 1 & \bar{2} & & \\
2 & 3 & 9 & 4 & 1 & 2 & 5 \\
 & \bar{9} & 3 & \bar{5} & 4 & \bar{3} & \\
\hline
2 & \bar{9} & & = & 11 & &
\end{array}
$$

Since 11 is not divisible by 17 the answer is no.

Example: Check whether 16409012 is divisible by 17.

We again use the negative Ekadhika of 17 which is 5.

$$
\begin{array}{cccccccc}
 & \bar{4} & & \bar{1} & 4 & & & \\
1 & 6 & 4 & 0 & 9 & 0 & 1 & 2 \\
 & 7 & \bar{1} & 9 & \bar{2} & 5 & 9 & \\
\hline
1 & 7 & & & & & &
\end{array}
$$

Thus, the given number is divisible by 17.

Example: Check whether 1645267 is divisible by 19.

The Ekadhika of 19 is 2 and it is preferable to use the positive Ekadhika to check divisibility.

```
   1  1     2
1  6  4  5  2  6  7
   9  1  3  4  0
-----
1  9
```

Since we obtain 19 the answer is yes.

Example : Check whether 1502383 is divisible by 23.

The positive Ekadhika of 23 is 7 and it is preferable to use this.

```
3  3  6  6  2
1  5  0  2  3  8  3
   6  4  4  8  9
----
4  6
```

Since 46 is divisible by 23, the answer is yes.

Example: Check whether 2342752 is divisible by 29.

The positive Ekadhika of 29 is 3 and it is preferable to use this.

```
1  2     1  1
2  3  4  2  7  5  2
   1  2  6  1  1
-----
3  1
```

Since 31 is not divisible by 29, the answer is no.

Example: Check whether 18345342 is divisible by 29.

We once again use the positive Ekadhika 3.

$$
\begin{array}{cccccccc}
1 & 2 & 2 & 1 & & 1 & & \\
1 & 8 & 3 & 4 & 5 & 3 & 4 & 2 \\
 & 9 & 3 & 6 & 7 & 4 & 0 & \\
\hline
2 & 9 & & & & & &
\end{array}
$$

Since we obtain 29 the answer is yes.

Example: Check whether 1205016 is divisible by 37.

The positive Ekadhika of 37 is 26 and the negative Ekadhika is 11. It is preferable to use the latter.

$$
\begin{array}{ccccccc}
10 & & \bar{9} & 4 & \bar{6} & & \\
1 & 2 & 0 & 5 & 0 & 1 & 6 \\
 & 1 & \bar{9} & 0 & 9 & \bar{5} & \\
\hline
11 & 1 & & & & &
\end{array}
$$

As 111 is divisible by 37 the answer is yes.

Example: Check whether 1921276 is divisible by 59.

The Ekadhika of 59 is 6 and it is preferable to use this.

$$
\begin{array}{ccccccc}
4 & 4 & 2 & 2 & 4 & & \\
1 & 9 & 2 & 1 & 2 & 7 & 6 \\
 & 9 & 6 & 7 & 4 & 3 & \\
\hline
5 & 9 & & & & &
\end{array}
$$

Since we obtain 59 the answer is yes.

Thus, it is easy to determine whether a number is divisible by other numbers or not. We had mentioned earlier that this technique is useful for checking the divisibility of a number by prime numbers. A little thought will indicate that the technique is also applicable to determine the divisibility by any number ending in 1, 3, 7 or 9 whether it is prime or not. For example, divisibility by 91 can be determined by using its negative Ekadhika 9 even though 91 itself is a product of 13 and 7. It should also be obvious that the technique described here is helpful in factoring numbers.

Exercises

1. Determine whether the following numbers are divisible,

a) 34214 by 2
b) 429322 by 3
c) 31428 by 3
d) 41324 by 4
e) 31252 by 5
f) 21356 by 6
g) 52344 by 6
h) 21494 by 8
i) 34288 by 8
j) 41238 by 9

2. Find the Ekadhika and negative Ekadhika of the following numbers.

a) 43
b) 47
c) 53
d) 41

e) 79
f) 149
g) 283
h) 359
i) 131
j) 101

3. **Determine whether the following numbers are divisible.**

a) 51234 by 7
b) 34296 by 11
c) 46081 by 7
d) 851214 by 13
e) 841754 by 23
f) 429362 by 39
g) 2550522 by 39
h) 429364 by 41
i) 1500518 by 41
j) 3257222 by 89

4. **Factorise the following numbers.**

a) 10164
b) 49725
c) 176341
d) 126469
e) 35929

Notes and Explanations

The mathematical methods presented in this book may seem astonishing, but are actually based on some simple principles of mathematics. The essential principles involved are the place-value system and some elementary algebra, over which Indians had mastery for a long time. The methods were therefore worked out long ago, and have been widely used by common people in India.

Here, we give some arguments justifying the methods. We use as little of algebra as possible. Detailed proofs have been avoided. However, these can be easily worked out by a reader based on the arguments presented here.

1. Carpenter's method for multiplication

For problems involving two different units, such as feet and inches, we can write the conversion factor as x. Thus a ft b in can be written as ax + b, where x in this case is 12. Multiplying 5x + 1 and 3x + 5 gives $(5 \times 3)x^2 + (5\times5 + 1\times3)x + 1\times5$.

The x^2 term, or the square-feet term, is obtained by multiplying the feet-part of the length and breadth. The x^0 term, or the square inch term, is obtained by multiplying the inch-part of the length and breadth. The cross-multiplication gives us $5\times5 + 1\times3$, which is the coefficent of x, having units of feet × inch. Dividing this by x gives the square-feet-part and multiplying the remainder by x gives the square-inch-part.

2. Navasesh or Modulo-9

Navasesh or modulo-9 of a number, is the remainder obtained when the number is divided by 9. We can write a number, say 3472 as follows :

$$\begin{aligned} 3472 &= 3\times10^3 + 4\times10^2 + 7\times10^1 + 2\times10^0 \\ &= 3\times(999 + 1) + 4\times(99 + 1) + 7\times(9 + 1) + 2 \\ &= (3\times999 + 4\times99 + 7\times9) + (3+ 4 + 7 + 2) \end{aligned}$$

Since 999, 99, 9 are multiples of 9, the term in the first bracket above is a multiple of 9. Therefore,

$$N(3472) = [3472]_9 = [3 + 4 + 7 + 2]_9$$

Thus Navasesh of a number can be obtained by adding the individual digits of the number. If the result is greater than 9, the digits are repeatedly added to obtain a single digit number.

3. Check using Navasesh

By the definition of Navasesh, we can write a number, say 4127, as $a\times9 + N(4127)$, where a is an appropriate integer. Thus, we obtain the following results.

(i) $4127 + 5143 = \{a\times 9 + \mathbf{N}(4127)\} + \{b\times 9 + \mathbf{N}(5143)\}$,
where a and b are appropriate integers,
$= (a + b) \times 9 + \mathbf{N}(4127) + \mathbf{N}(5143)$.

Therefore,

$$\mathbf{N}(4127 + 5143) = \mathbf{N}[\mathbf{N}(4127) + \mathbf{N}(5143)].$$

(ii) $6234 - 2135 = \{c \times 9 + \mathbf{N}(6234)\} - \{d \times 9 + \mathbf{N}(2135)\}$,
where c and d are appropriate integers,
$= \{(c - d) \times 9\} + \{\mathbf{N}(6234) - \mathbf{N}(2135)\}$.

Therefore,

$$\mathbf{N}(6234 - 2135) = \mathbf{N}[\mathbf{N}(6234) - \mathbf{N}(2135)].$$

(iii)
$$\begin{aligned} 5429 \times 3147 &= \{e \times 9 + \mathbf{N}(5429)\} \times \{f \times 9 + \mathbf{N}(3147)\} \\ &= \{e\times f \times 9 + e \times \mathbf{N}(3147) + f \times \mathbf{N}(5429)\} \times 9 \\ &\quad + \mathbf{N}(5429) \times \mathbf{N}(3147). \end{aligned}$$

Therefore,

$$\mathbf{N}(5429) \times \mathbf{N}(3147) = \mathbf{N}[\mathbf{N}(5429) \times \mathbf{N}(3147)]$$

4. Nikhilam

Now, let us consider the multiplication of 93 and 91,

$93 = 100 - 07, \qquad 91 = 100 - 09$

$$\begin{aligned} 93 \times 91 &= (100 - 07)\times (100 - 09) \\ &= 100 \times 100 - 100\,(7 + 9) + 7\times 9 \\ &= 100(100 - 7 - 9) + 7\times 9 \\ &= 100(93 - 9) + 7 \times 9 \\ \text{or} \quad &= 100(91 - 7) + 7 \times 9. \end{aligned}$$

Note that 7×9 is the multiplication of the two deviations. The first term is 100 times the cross-subtraction.

5. Nikhilam for three-number multiplication

Consider now the multiplication of 97, 105 and 91.

$$97 \times 105 \times 91 = (100 - 3) \times (100 + 5) \times (100 - 9)$$

$$= 100 \times 100 \times 100 + 100 \times 100 \times (\bar{3} + 5 + \bar{9}) + 100 \times (\bar{3} \times 5 + \bar{3} \times \bar{9} + 5 \times \bar{9}) + \bar{3} \times 5 \times \bar{9}$$

$$= \{100 \times 100[100 + \bar{3} + 5 + \bar{9}]\} + \{100 \times (\bar{3} \times 5 + 5 \times \bar{9} + \bar{3} \times \bar{9})\} + \{\bar{3} \times 5 \times \bar{9}\}.$$

The third term is the Nikhilam term for units place, the second term is the Nikhilam term for hundreds place and the first term is 97 + 5 – 9, or 105 – 3 – 9, or 91 – 3 + 5, the Nikhilam term at ten-thousands place.

6. Nikhilam corollary for special multiplication

If we want to multiply 253 and 257, we note that the unit-place digits of the two numbers add to 10 and the higher digits are the same. Therefore,

$$\begin{aligned} 253 \times 257 &= (250 + 3) \times (250 + 7) \\ &= 250 \times 250 + 250 \times (3 + 7) + 3 \times 7 \\ &= 250 \times (250 + 3 + 7) + (3 \times 7) \\ &= (25 \times 26 \times 100) + (3 \times 7). \end{aligned}$$

The term in the second bracket above is a multiplication of the digits at units place, whereas the first term is 25 × (25+1) times hundred.

7. Urdhva Tiryaka

Let us carry out long multiplication of numbers $a_3\ a_2\ a_1\ a_0$ and $b_3\ b_2\ b_1\ b_0$ where a_i's and b_i's are decimal digits.

			b_3	b_2	b_1	b_0
		$\times$	a_3	a_2	a_1	a_0
			a_0b_3	a_0b_2	a_0b_1	a_0b_0
		a_1b_3	a_1b_2	a_1b_1	a_1b_0	
	a_2b_3	a_2b_2	a_2b_1	a_2b_0		
a_3b_3	a_3b_2	a_3b_1	a_3b_0			
a_3b_3	$a_2b_3 + a_3b_2$	$a_1b_3 + a_2b_2 + a_3b_1$	$a_0b_3 + a_1b_2 + a_2b_1 + a_3b_0$	$a_0b_2 + a_1b_1 + a_2b_0$	$a_0b_1 + a_1b_0$	a_0b_0

In Urdhva Tiryaka, the $a_3\ a_2\ a_1\ a_0$ is reversed as $a_0\ a_1\ a_2\ a_3$ and placed below $b_3\ b_2\ b_1\ b_0$ with a_0 below b_0. The overlapping numbers are multiplied to yield the right-most term as a_0b_0. The digits $a_0\ a_1\ a_2\ a_3$ are now shifted one digit left and the overlapping digits are multiplied to get the second term from the right, i.e., $a_0b_1 + a_1b_0$. Shifting again to the right and multiplying the overlapping digits, yields the third term as $a_0b_2 + a_1b_1 + a_2b_0$. Similarly the fourth term is obtained as $a_0b_3 + a_1b_2 + a_2b_1 + a_3b_0$, and so on. The results are the same as that obtained by long multiplication.

8. Squares

The square of a number $a_3\ a_2\ a_1\ a_0$ can be written as follows :

			a_3	a_2	a_1	a_0
a_3^2	$2a_3a_2$	$2a_3a_1 + a_2^2$	$2a_3a_0 + 2a_2a_1$	$2a_2a_0 + a_1^2$	$2a_1a_0$	a_0^2

It is obvious that the right–most term is **D**(a_0). The next term to the left is **D**(a_1a_0). The third term is **D**$(a_2a_1a_0)$, the fourth term is **D**$(a_3a_2a_1a_0)$, the fifth term is **D**$(a_3a_2a_1)$ and so on. The square of a number can therefore be easily obtained using Dwandwa.

9. Straight-Division

Let us take a division example where the dividend, divisor, quotient and remainder are expressed as polynomials P(x), D(x), Q(x) and R(x), respectively, as follows :

$$P(x) = b_6x^6 + b_5x^5 + b_4x^4 + b_3x^3 + b_2x^2 + b_1x^1 + b_0x^0$$

$$D(x) = a_3x^3 + a_2x^2 + a_1x^1 + a_0x^0$$

$$Q(x) = q_3x^3 + q_2x^2 + q_1x^1 + q_0x^0$$

$$R(x) = R_3x^3 + R_2x^2 + R_1x^1 + R_0x^0$$

Let us carry out the long division as illustrated below, where S_i and I are intermediate results.

$a_3\,a_2\,a_1\,a_0$)	b_6	b_5	b_4	b_3	b_2	b_1	b_0
	a_3q_3	a_2q_3	a_1q_3	a_0q_3			
	S_3	I	I	I	b_2		
		a_3q_2	a_2q_2	a_1q_2	a_0q_2		
	0	S_2	I	I	I	b_1	
			a_3q_1	a_2q_1	a_1q_1	a_0q_1	
		0	S_1	I	I	I	b_0
				a_3q_0	a_2q_0	a_1q_0	a_0q_0
				R_3	R_2	R_1	R_0

The quotient and Remainder digits in the above example are determined as follows :

q_2 : $\{S_3 \times 10 + b_5\} - a_2q_3 = a_3q_2 + S_2$

q_1 : $\{S_2 \times 10 + b_4\} - a_1q_3 - a_2q_2 = a_3q_1 + S_1$

q_0 : $\{S_1 \times 10 + b_3\} - a_0q_3 - a_1q_2 - a_2q_1 = a_3q_0 + R_3$

R_2 : $b_2 - a_0q_2 - a_1q_1 - a_2q_0 = R_2$

R_1 : $b_1 - a_0q_1 - a_1q_0 = R_1$

R_0 : $b_0 - a_0q_0 = R_0$

Note the terms involved in determining any of the quotient or remainder digits. This is precisely what is done in Straight-Division. Straight-Division is therefore in-place computation to determine one quotient digit at a time.

The terms involved in go-back can be similarly shown to be the same as what we used in straight-division.

10. Square Root

We had earlier described a method to find the square of a number using Dwandwa. We computed the digits of the square from right to left. The technique is equally applicable if we do so from left to right. We illustrate this to determine the square of $a_1 . a_2\ a_3\ a_4$.

a_1 .	a_2	a_3	a_4			
$D(a_1)$.	$D(a_1a_2)$	$D(a_1\,a_2\,a_3)$	$D(a_1\,a_2\,a_3\,a_4)$	$D(a_2\,a_3\,a_4)$	$D(a_3\,a_4)$	$D(a_4)$

Also note that,

$$D(a_1a_2a_3a_4) = 2a_1a_4 + D(a_2a_3)$$

$$D(a_1a_2a_3) = 2a_1a_3 + D(a_2)$$

$$D(a_1a_2) = 2a_1a_2.$$

Let us try to determine the square root of $P=p_0\ p_1 .\ p_2\ p_3\ p_4\ p_5$. The square root will be represented by $a_1 .\ a_2\ a_3\ a_4$. The first digit a_1 of the square root is determined as the largest digit whose dwandwa or square is less than (or equal to) p_0p_1 We therefore have $p_0p_1 - D(a_1) = I_1$, where I_1 is the appropriate intermediate remainder. We will use $2a_1$ as the divisor, d. Since the second digit in the square of $a_1 .\ a_2a_3...$ can be written as $D(a_1a_2)$ or $2a_1a_2$, we can therefore determine a_2 by dividing $I_1x + p_2$ by d as follows :

$$I_1\,x + p_2 = d\,a_2 + I_2$$

Now the third digit in the square of $a_1 .\ a_2\ a_3$ can be written as $D(a_1a_2a_3)$ and therefore, a_3 can be obtained by dividing $I_2\,x + p_3 - D(a_2)$ by d as follows :

$$l_2 x + p_3 - D(a_2) = d\, a_3 + l_3.$$

Similarly a_4 and a_5 can be obtained by

$$l_3 x + p_4 - D(a_2 a_3) = d\, a_4 + l_4$$

and $$l_4 x + p_5 - D(a_2 a_3 a_4) = d\, a_5 + l_5.$$

This is precisely the technique used for determining square root.

11. Divisibility

Let us determine the divisibility of 743528 by 7. The Ekadhika of 7 is determined as 5. We can write,

$$743528 = 74352 \times 10 + 8,$$
where 10 is the base and $5 \times 10 = 7 \times 7 + 1$.

So,

$$\begin{aligned} 743528 \times 5 &= 74352 \times 5 \times 10 + 5 \times 8 \\ &= 74352 \times (7 \times 7 + 1) + 5 \times 8 \\ &= 74352 \times 7 \times 7 + (74352 + 5 \times 8) \end{aligned}$$

If 743528 is divisible by 7, then so is $743528 \times 5 - 74352 \times 7 \times 7$, which is equal to $74352 + 5 \times 8$. This is the process of Vetsana.

Similarly the negative Ekadhika can be used. Let us find whether a number 352943 is divisible by 13. The negative Ekadhika of 13 is 9. We can write,

$$352943 = 35294 \times 10 + 3 \text{ and } 9 \times 10 = 13 \times 7 - 1.$$

Therefore,

$$\begin{aligned}352943 \times 9 &= 35294 \times 9 \times 10 + 3 \times 9\\ &= 35294 \times (13 + 7 - 1) + 3 \times 9\\ &= 35294 \times 13 \times 7 - \{35294 - 3 \times 9\}.\end{aligned}$$

If 35294 is divisible by 13, then so is $35294 - 3 \times 9$. This is the Vetsana defined earlier.

Index